Industrial
Pollution Control:
Issues and Techniques

Second Edition

Related Titles of Interest from Van Nostrand Reinhold

AIR POLLUTION ENGINEERING MANUAL
by the Air Waste Management Association

PROSPERITY WITHOUT POLLUTION
by Joel Hirschorn

CONTINUOUS EMISSIONS MONITORING
by Jim Jhanke

AIR POLLUTION MODELLING
by Paolo Zannetti

Industrial Pollution Control:

Issues and Techniques

Second Edition

by Nancy J. Sell

VNR VAN NOSTRAND REINHOLD
_____ New York

Copyright © 1992 by Van Nostrand Reinhold

Library of Congress Catalog Card Number 91-45496
ISBN 0-442-00658-6

I(T)P   Van Nostrand Reinhold is an International Thomson Publishing company.
      ITP logo is a trademark under license.

Printed in the United States of America

Van Nostrand Reinhold   International Thomson Publishing GmbH
115 Fifth Avenue   Konigswinterer Str. 518
New York, NY 10003   5300 Bonn 3
                                   Germany

International Thomson Publishing   International Thomson Publishing Asia
Berkshire House,168-173   38 Kim Tian Rd., #0105
High Holborn, London WC1V 7AA   Kim Tian Plaza
England   Singapore 0316

Thomas Nelson Australia   International Thomson Publishing Japan
102 Dodds Street   Kyowa Building, 3F
South Melbourne 3205   2-2-1 Hirakawacho
Victoria, Australia   Chiyada-Ku, Tokyo 102
                                   Japan

Nelson Canada
1120 Birchmount Road
Scarborough, Ontario
M1K 5G4, Canada

16 15 14 13 12 11 10 9 8 7 6 5 4 3 2

Library of Congress Cataloging-in-Publication Data
Sell, Nancy Jean, 1945-
   Industrial pollution control / by Nancy J. Sell.—2nd ed.
      p. cm.
   Includes bibliographical references and index.
   ISBN 0-442-00658-6
   1. Factory and trade waste. 2. Industry—Environmental aspects.
I. Title.
TD897.5.S45   1992
628.4—dc20   92-19145
                                      CIP

For my mother, Haydn, Hans, and Joie

Contents

Preface

Since the first edition of this book was published, many things related to pollution and pollution control have changed—some for the better, but some for the worse. In many ways it is a time of contradictions. People are becoming enthusiastic about environmental issues. Recycling, in particular, has been avidly embraced by a large percentage of the general population. Better pollution control methods are being developed and, even more importantly, more efficient industrial processing techniques are available that result in less pollution and fewer wastes. On the other hand, never before has such massive man-made pollution occurred, as in Kuwait in February 1991, with the torching of hundreds of oil wells and the intentional spillage of millions of barrels of oil. Stories abound of severe industrial pollution problems in the former Eastern Block countries. And we are discovering more and more types of problems of which we were never before aware.

Scientists continue to debate the effects of many known and suspected pollutants. It is now suspected that dioxin and other toxics were given a "bum rap." Some scientists have predicted extensive global effects from sulfur and carbon dioxide emissions, but even the burning of the Kuwaiti high sulfur crude oil has not yet led to more than just regional problems. Only time and, in some cases, more scientific studies will sort out the real effects of these substances. Meanwhile, in this edition, as in the first, I have attempted to take a neutral stance on the various environmental issues, considering both the known or suspected problems, but also the successes and the questions.

Some of the chapters have been extensively modified, largely to consider new technologies, new regulations, and newly discovered environmental issues. Those greatly expanded include the chapters on the pulp and paper industry (chapter 12) and the chemical industries (chapter 16). Chapter 17, regarding the overall environmental and economic impact of the EPA and related regulations, has been totally updated. In many ways, the status of the environment and the environmental regulations themselves are quite different in the United States than they were 11 years ago! A large number of the photographs were also updated and expanded in this edition. Many were replaced that did not illustrate some of the problems, processes, or control techniques well, and a number were added that potentially increase the understanding of the material.

I sincerely thank the many companies and governmental agencies that have provided information, photographs, and in some cases even edited what I wrote about them, for this edition as well as the first. Some chose not to be referenced. Without their help, however, much of the information would not have been available.

Nancy Sell

1

The Nature of Pollution

INTRODUCTION

Industries have, since the start of the industrial revolution, been plagued with pollution problems of all types. Only since the early 1970s, however, has there been any major effort by the federal government to clean up our environment, and thus to force industries to meet stringent emission standards.

How have the industries met this challenge? What methods are they using to control their emissions? To answer these questions it is first desirable to consider the nature of various types of pollution, some of their effects on the environment, and the current regulatory trends. With this background, we can first consider the general methods of pollution control, and then examine some particular types of industries in detail: the particular chemical and/or physical processes used, how and where the pollution is generated, and what methods are typically used to control those emissions. The final chapter considers the overall impact of the pollution regulations on both our environment and on industry.

AIR POLLUTION

Air pollution can be defined as the addition to our atmosphere of any material(s) having a deleterious effect on life. Typical air pollutants include things such as carbon monoxide (CO), nitrous oxides (NO_x), sulfur oxides (SO_x), and various hydrocarbons and particulates (small particles, usually of solids). It can be one of two types:

1. A primary pollutant, or one lethal as it originates from the source; or
2. A secondary pollutant, one formed through the reaction of primary pollutants. (This reaction can occur at the emission point or at far removed localities.)

Smog, as illustrated in Figure 1-1, is comprised primarily of secondary pollutants.

Air pollution is generated by six major types of sources:

1. Transportation
2. Domestic heating
3. Electric power generating
4. Refuse burning
5. Forest and agricultural fires
6. Industrial fuel burning and process emissions.

Our primary task is, of course, to focus on the last type, the industrial sources, which contribute about 20% of our total air pollution. Also of some concern is the pollution formed during electric power generation. Many industries consume tremendous amounts of electricity. This electricity may be bought from a power company or generated on site. Either way, the industry is responsible for the environmental problems that arise due to its use.

Considering only industrial sources, Figure 1-2 illustrates the relative percentages of each type of pollutant. Further consideration of various types of industry allows one to rank those with the greatest potential emission (Table 1-1).

When discussing the effects of these pollutants, it is necessary to examine the effects on animals (including man), on plants, and on materials, as well as the

FIGURE 1-1.   Manhattan in a shroud of smog. (*Photo courtesy DOE.*)

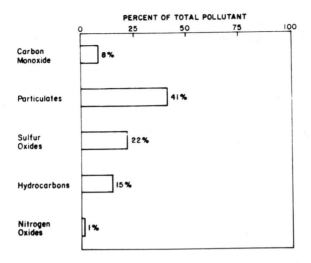

FIGURE 1-2.   Industrial air pollution. (Graph courtesy DOE; data from *Environmental Quality,* Council on Environmental Quality, p. 212.)

TABLE 1-1   Industrial Air Pollutant Emissions[a]

Type of Industry	Annual Emissions Levels (billion lb)	Types of Emissions
1. Petroleum refining	8.4	Particulates, sulfur oxides, hydrocarbons, CO.
2. Smelters for Al, Cu, Pb, Zn	8.3	Particulates, sulfur oxides.
3. Iron foundries	7.4	Particulates, CO.
4. Kraft pulp and paper mills	6.6	Particulates, CO, sulfur oxides.
5. Coal cleaning and refuse	4.7	Particulates, CO, sulfur oxides.
6. Coke (for steel manufacturing)	4.4	Particulates, CO, sulfur oxides.
7. Iron and steel mills	3.6	Particulates, CO.
8. Grain mills and grain handling	2.2	Particulates.
9. Cement manufacturing	1.7	Particulates
10. Phosphate fertilizer plants	0.624	Particulates, fluorides.

[a]Data from Ross, R.D., *Air Pollution and Industry.* New York: Van Nostrand Reinhold, 1972.

meteorological effects. These can range from nuisance and aesthetic damage, such as odor and low visibility, to property and/or health damage, and major local or even global ecosystem disruption, such as alteration of the climate. The precise effects depend on many factors, including climate, population density, and other local conditions. In spite of this variability, there are some general effects that have been documented. Though many substances can be air pollutants, this chapter will concentrate on some of the most common: carbon monoxide (CO), the sulfur oxides, the nitrogen oxides and related compounds, hydrocarbons, ozone (O_3), and particulates.

Effects on Man

Health Effects

Many common air pollutants can have very serious effects on human health. For example, CO is known to contribute to heart disease. Considering all sources of pollution, especially transportation, it is one of the major, if not *the* major, pollutants, and is also one of the most difficult to eliminate. It is formed during the combustion of carbon-containing compounds whenever there is a lack of oxygen (O_2): $2C + O_2 \rightarrow 2\,CO$ in limited O_2; $C + O_2 \rightarrow CO_2$ in excess O_2. Though industry is not, in general, a large producer of CO, it can, in some instances, contribute to the overall CO problem. This substance affects the central nervous system, even in very low concentrations, by forming carboxyhemoglobin in the bloodstream, which interferes with the normal transport of O_2 to the body cells. Two percent carboxyhemoglobin is enough to generate observable effects, and can be formed by an 8-hour exposure to only 10 ppm CO, only slightly over the ambient U.S. air quality standard of 9 ppm for an 8-hour exposure. O_2 transport is clearly affected at 75% carboxyhemoglobin, generated at CO levels of 30 ppm or greater. This is serious when one considers that heavy automobile traffic can generate 50–140 ppm CO, and the smoking of cigarettes can create up to 15% carboxyhemoglobin. Thus, New York cab drivers who smoke can be in serious health trouble. In addition to the immediate effects of poor oxygen transport, high carboxyhemoglobin levels tend to make a person retain cholesterol in the aorta.

The sulfur oxides SO_2 and SO_3 are generated primarily in the combustion of high-sulfur fuels. Some industrial processes, such as copper smelting and paper pulping, also generate large amounts of the sulfur oxides. The normal exposure threshold is 5 ppm for an 8-hour period. Most industries do not exceed this, at least at ground level, if they have tall stacks. Smelters are one exception to this in that they can attain 30–40 times the threshold level for short periods of time, particularly if there is a temperature inversion. A temperature inversion occurs when a cooler lower layer of air is trapped by a warmer upper layer of air. The pollutants build up in this cooler lower layer because the cool air cannot rise and carry them away.

The sulfur oxides are toxic to the human body, especially if the person has a previous respiratory disease such as emphysema and/or is older. They can also accentuate viral pneumonia. The sulfur oxides usually can be detected by their odor, but prolonged exposure may desensitize a person to these compounds.

The nitrous oxides NO and NO_2 are usually found in much lower concentrations. They are generated only in high-temperature combustion situations, and hence have been referred to as an elitist pollutant, only present in technologically advanced societies. Their ultimate effect on humans is not clearly understood, but they do act as irritants to breathing, and create discomfort to the eyes. NO_2 can also destroy the celia in the respiratory system and suppress alverolae macrophage activity, the lungs' final defense against foreign matter. There is evidence that the nitrogen oxides can have very serious consequences. High concentrations of NO_2 do form in farm silos, creating "silage gas" during aerobic decay of the silage, and there are many documented instances of farmers being fatally overcome by these fumes. Concentrations this large are not likely to form, fortunately, due to normal industrial operations.

Recent studies of various nitrogeneous air pollutants have indicated that these compounds may be more serious health hazards than once thought. In particular, the peroxynitrates are quite stable at lower air temperatures, and may be more important pollutants than O_3 in northern cities such as New York and Chicago. Peroxyacetylnitrate (PAN) is created by photochemical reactions involving hydro-carbon. PAN has a general structure of
$$R-\overset{\overset{\displaystyle O}{\|}}{C}-O-O-N-O,$$
where R stands for hydrocarbon chains of varying lengths ($CH_3-CH_3\ CH_2-...$). A typical formation mechanism would be as follows:

$$R-\overset{\overset{\displaystyle O}{\|}}{C}-R \xrightarrow{UV} R\cdot + R-\overset{\overset{\displaystyle O}{\|}}{C}\cdot$$
a hydrocarbon
(a ketone)

$$R-\overset{\overset{\displaystyle O}{\|}}{C}\cdot + O_2 \rightarrow R-\overset{\overset{\displaystyle O}{\|}}{C}-O-O\cdot$$

$$R-\overset{\overset{\displaystyle O}{\|}}{C}-O-O\cdot + NO_2 \rightarrow R-\overset{\overset{\displaystyle O}{\|}}{C}-O-O-\overset{\overset{\displaystyle O}{|}}{N}-O\cdot$$
(PAN)

Some PAN is generated naturally, for coniferous vegetation is a major source of hydrocarbons and NO_2 is prevalent everywhere. NO can also react with some polycyclic aromatic hydrocarbons in laboratory tests to produce mutagenic compounds. Though mice and rats can be protected somewhat from the effects of NO by large doses of Vitamin E, no such effect is possible for humans.[1]

There are hundreds of types of hydrocarbons that form air pollutants. Many of them are possibly carcinogenic and might be at least partially responsible for the current increase in lung cancer.

Ozone, O_3, is usually generated by the photochemical reaction between the various nitrogen oxides and O_2 in the atmosphere. For example, $NO_2 \xrightarrow{\text{sun}} NO + O$. The $O \cdot$ is a free radical, an oxygen atom with seven electrons, one unpaired. Then $O \cdot + O_2 \rightarrow O_3$ Ozone aggravates asthma, emphysema, and chronic bronchitis. It lowers one's resistance to infections, reduces the ability to concentrate, and induces coughing, chest discomfort, and irritation of the upper respiratory system. It even causes chromosome damage in laboratory animals, and has been linked to a type of anemia in humans.[2]

Particulates have various adverse effects, dependent upon their size:

1. Below 0.1 μm, the major effects relate to weather modification. This is the most likely size to induce nucleation of water droplets.
2. Between 0.4 and 0.8 μm, the diameters are approximately equal to the wavelength of light, and thus lead to the greatest restriction of visibility.
3. Between 1 and 5 μm, there is maximum deposition in the lungs upon inhalation.
4. Between 3 and 15 μm, the particulates are deposited in the upper respiratory system.
5. Between 10 and 100 μm, the particulates create dustfall and dirt.

Though the categories overlap, and particles of any size may show any of these effects to some extent, this is a useful breakdown when considering the major effects.[3,4]

Those particulates that are inhaled are damaging to respiratory systems. In addition, they may be toxic. For example, mercury and other heavy metals lead to direct biochemical reactions. Most, however, are inorganic and nontoxic substances. When ingested through the nose and mouth, along with the approximately 7600 liters of air we inhale daily, the particulates may end up deposited in the lungs, causing a buildup on the lung lining. This could result in a disease called silicosis. It occurs particularly frequently in those individuals involved in mining and cement manufacture. Black lung disease, a form of silicosis, has been commonly found in coal miners. This buildup on the lungs reduces the ability of the lungs to transfer O_2 into the blood. The normally elastic and spongy lung tissues harden, reducing the lung's breathing efficiency. In order to pump an adequate O_2 supply, the heart must then work harder. This leads to shortness of breath, possibly to an enlarged heart, and eventually to premature death.

In addition, particulates can sometimes cause excessive mucus secretion as a protective reflex. This excess mucus can restrict the bronchiole tubes and lead to bronchitis. For other individuals, the pollutants could cause enough irritation of the bronchial tubes for them to constrict, increasing one's chances of developing emphysema.

Though most of the detrimental health effects are caused by the inhaled particulates that are small enough to penetrate deep into the lungs, recent evidence has shown that those deposited in the upper respiratory system can also lead to health problems.

There are many public interest groups that are unwilling to compromise when considering human health, with much justification. There have been situations when large amounts of pollutants, combined with particular meteorological conditions, have led to large numbers of directly attributable deaths: for example, in London, in 1952 (3500–4000 deaths) and 1956 (900 deaths); in Donora, Pennsylvania, in 1948 (20 deaths); and in New York, in 1956 (400 deaths). A study by T. A. Hodgson, Jr., of 2 ½ years of New York City data led him to report that even slight or moderate increases in the concentration of air pollution (increases that may not even be noticed) can be expected to result in an increased mortality from heart and respiratory diseases amounting to several hundred deaths per month.[5]

There are data to suggest that the worst condition for human health is the combination of particulates with a high SO_2 concentration. A very large percentage of the SO_2 in our atmosphere is due directly or indirectly to natural sources (volcanoes, decaying vegetation, sea spray, etc.). The SO_2 generated naturally is, however, so dispersed over the world that it never builds up to dangerous levels. Man's SO_2 contribution-the "anthropogenic" SO_2-tends to be concentrated in industrial and urban areas, and hence can rise to dangerous levels. The SO_2 in the air, often with particulates acting as a catalyst, can be converted to SO_3:

$$2SO_2 + O_2 \xrightarrow{\text{particulates}} 2SO_3.$$

The particulate surfaces can provide a reaction site for the formation of SO_3 to occur. The SO_3 can readily react with water vapor to produce sulfuric acid (H_2SO_4). This acid can easily damage lung tissues. This particular problem is often referred to as industrial (or gray) smog and it predominated in the older cities that depended heavily on the use of coal and oil.

In the atmosphere, some of the sulfuric acid droplets can react with ammonia (NH_3) to generate solid ammonium sulfate, $(NH_4)_2SO_4$. Donora, Pennsylvania, in 1948, when it experienced the large numbers of deaths, had high atmospheric concentrations of acid sulfate salts (principally zinc ammonium sulfate and zinc sulfate), and those are thought to be the chief contributors to the disaster.

The H_2SO_4 in the air gets washed down whenever it rains, generating "acid rain," a phenomenon that has been rising rapidly since the early 1960s. Acid rain has been linked not only to damaged trees and other plants, increased weathering and corrosion of materials and buildings, and water pollution problems, but also is possibly an added threat to human health.[6] Acid rain is also formed from NO emissions, which can react to form nitric acid (HNO_3) in the atmosphere. In the late 1970s, acid rain was declared as possibly "the most severe environmental problem of the century."[7]

Visibility

Visibility (or the lack of it) has a psychological impact on man. On a clear day in a pollution-free environment, one can see approximately 163 miles.[8] As the air becomes more polluted, the visibility becomes poorer and poorer.

The decrease in visibility is caused by a variety of pollutants. The small particulates (0.1-1 μm in diameter) scatter and absorb light. Sulfate compounds, carbonaceous soot, and nitrate compounds commonly form particulates in this size range. NO_2, a gas, absorbs blue light, causing the air to appear yellow or brownish red. Copper smelters, fossil fuel–generating plants, and automobiles are some of the chief culprits in reducing visibility.

FIGURE 1-3.   A temperature inversion can, in particular, maintain smog at a particular atmospheric level, as illustrated here above Phoenix, Arizona. (Photo courtesy of David Von Oepen.)

The visibility is a somewhat subjective problem that can be quantified by the "extinction coefficient," a measure of the sum of the scattering and absorption of light by both fine particulates and gases. The extinction coefficient is a measure of the light lost in traveling through the atmosphere; hence, the greater the extinction coefficient, the worse the visibility. As long as the weather is fairly dry (below 70% relative humidity), the extinction coefficient is directly proportional to the concentration of appropriately sized particulates.

Currently, the average visibility is bad, and getting worse, in the Northeast. The median visibility is only 4-12 miles in metropolitan areas and 9-14 miles in nonurban areas. In the Southwest, on the other hand, visibility ranges from 30-35 miles in urban areas and 65-80 miles or greater in nonurban locations.[9]

Odor

Odor, like visibility, has primarily a psychological impact on man. Odor—and taste—are strictly subjective senses. A person's nose is still the best odor-measuring device, though it does, of course, present difficulties when one attempts to quantify the extent of the odor. How can one judge whether an industrial odor is acceptable or not?

The ASTM guidelines, also the basis for current EPA guidelines, involve the selection of a panel of eight or more judges. These judges, preferentially female since women reportedly have a keener sense of smell than men, must not smoke nor be suffering from a cold. The test consists of having 100-ml samples of air being injected into each judge's nostril with a syringe. If an odor is detected, that air sample is diluted with clean air until no further odor is detectable. That dilution level is called the "odor threshold." The odors are measured in "odor units" consisting of that quantity of odor that, when dispersed in one 1 ft^3 of odor-free air, is just detectable by 50% of the judges. If, for example, a sample was diluted 200 times to reach the threshold, that sample contains 200 odor units/ft^3.

There are other tests that can be used to "measure" odor. These include gas flow rates and qualitative and quantitative determination of odorous compounds by means of instruments such as gas chromatographs and mass spectrometers. None of these methods, however, really determines odor, and they are not used when considering air pollution measurements.

Effects on Animals

The health effects of the various pollutants on animals are much the same as their effects on humans. In addition, insecticides may also be a major problem for animals. Frequently, their food sources become contaminated by one form or another of air pollutant, and there are numerous documented cases of large numbers of animal deaths that can be directly related to this problem. Many pesticides can be carried right through the food chain. For example, if a pesticide is sprayed over a large area, much of it ends up in a lake or stream for consumption by fish. The fish can be eaten by certain types of birds, which can then be affected. Chlorine-containing pesticides have, for instance, been related to thinner than normal eggshells in fish-consuming predatory birds. The thin eggshells lead to breakage before the eggs hatch, and hence to a loss of those offspring.

Effects on Plants

Annually, the agricultural losses due to air pollution run in the hundreds of millions of dollars, much of the damage occurring, perhaps not surprisingly, in California. Damage can range from simply visible damage to that affecting growth and productivity and eventually destroying the life processes. Each pollutant or combination of pollutants creates its own type of damage, and from an examination of the plant it is often possible to identify the nature of the pollutant.[10]

The major pollutants that affect plant life are the primary pollutants SO_2 and hydrogen fluoride (HF), and the secondary pollutants O_3 and PAN.

SO_2 can have either chronic or acute effects on plant life. An initial bleaching of plant cells and a stunting of growth often leads to death. Smelters, one of the major industrial sources of SO_2, usually were surrounded by little, if any, plant life.

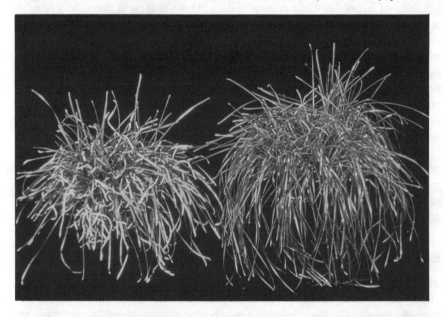

FIGURE 1-4.   Effect of pollutants on plants. Sometimes it is difficult to differentiate between the effects of pollutants and lack of proper nutrients. Illustrated here is bluegrass. That on the left has ozone damage, while the one on the right is normal. (Photo courtesy of the U.S. Department of Agriculture, Slide Set A-58–July 1972.)

In the Middle Atlantic states, potatoes, tomatoes, peas, corn, apples, and peaches are frequently damaged by SO_2.

Fluorides, though not a common air pollutant, do seriously affect plants. Fluorides usually are generated in the manufacture of phosphate fertilizers, and also in the reduction of aluminum ore. The usual symptom is that the leaf tissues die at the tip.

O_3, like SO_2, bleaches plant cells; the coniferous plants are particularly susceptible. Initially, the damage will show at the end of the leaf.

PAN also tends to be very reactive toward the nitrogen in plant materials, probably disrupting the bonds in protein molecules. The PAN and O_3 generated in California due to the intense automobile traffic and unfortunate geography (and hence climatory conditions), do significant damage at times to the Southern California citrus groves.

Particulates usually lead to phytotoxicity, inhibition of respiration, and/or photosynthesis. Typical culprits are cement kiln dust, soot, foundry dusts, and magnesium oxide. These dusts can thickly and completely cover the leaves, and not allow the normal biological activities to function.

Effects on Materials

One of the most serious problems affecting materials is due to the SO_2 and H_2SO_4 in the air—the acid rain. Calcium carbonate ($CaCO_3$), the primary constituent of limestone, is also a major ingredient in marble, mortar, and concrete. Unfortunately, the $CaCO_3$ reacts readily with H_2SO_4 to produce water-soluble calcium sulfate ($CaSO_4$):

$$CaCO_3 + H_2SO_4 \rightarrow CaSO_4 + CO_2 + H_2O.$$

The $CaSO_4$ is washed away in the rain, leaving the monument, statue, or building pitted, and exposing new $CaCO_3$ surfaces for further reaction. Over the past 40–50 years, more damage has been done to the ancient Greek and Roman remains than had been done for the previous 2000 years. Some of the historic early colonial headstones located in our northeastern cemeteries (the study of which can act as a technique for investigating the nature of the people of that time) are now becoming illegible.

The H_2SO_4 also reacts with materials such as metals, rubber, plastics, and some types of fabrics, partially or totally ruining them.

NO_2 can bleach fabrics, something particularly significant for some of the earlier synthetic dyes. It can also cause degradation of many fabrics. There have been several instances when the appropriate meteorological conditions combined with excess atmospheric NO_2 have led to the total disintegration of women's nylon

FIGURE 1-5.   Acid rain affects limestone, marble, mortar, and similar minerals, such as this figure on Chicago's Museum of Natural History. Sometimes the deterioration is actually occurring greater under the blackened areas, consisting of gypium and soot, for the rain cannot penetrate beneath to remove the acids as easily.

stockings (while they were wearing them). In addition, NO_2 aids in the corrosion of various nickel-brass alloys.

Particulates blacken and damage materials ranging from clothing to buildings. Many of our buildings are just now getting a sandblasting "facewash" after many years of exposure to the particulate contaminants in our cities.

Meteorological Effects and Changes

Air pollution can have a major effect on the climate, both regionally and globally. Regionally, rainfall can be drastically altered by the presence of air pollution. For example, La Porte, Indiana, located 30 miles southeast of the major steel center of Gary, Indiana, and also southeast of Chicago, averaged 47% more precipitation than the localities upwind of Chicago. In addition, as Chicago's haze and smog varies with the area's fuel use, so does La Porte's precipitation.[11]

The reason for this behavior can be seen if we consider how and why rain forms. The moisture in the air collects in the tiny droplets, using a particulate as the nucleus. These tiny droplets initially collect and form clouds. If a sufficient concentration of moisture is present, the droplets can attract more water vapor to themselves and grow in size and eventually form rain. The presence of the particulate thus catalyzes the initial moisture condensation and droplet formation.

Because of this behavior, air pollution can also have the opposite influence on precipitation. Too many particulates can encourage the formation of too many small, nuclear particles compared to the available moisture. Each particle cannot attract enough water vapor to itself, so it cannot grow enough to form rain droplets. The net effect is a decrease in precipitation.

The global climatic effects relate to the earth's heat balance. There are two proposed but opposing theories to predict the net effect of air pollution on the temperature:

1. *The greenhouse effect.* The increased burning of fossil fuels increases the amount of atmospheric CO_2. This CO_2 is important in that it is nearly transparent to visible radiation, but is a strong absorber and back radiator of infrared (heat) radiation. In this way, the atmospheric CO_2 acts as does the glass in a greenhouse. The visible light penetrates the CO_2 layer and is degraded on the earth's surface, and the heat formed cannot radiate back outward. Many large cities and metropolitan areas have become several degrees warmer, on the average, in recent years, and tend to be warmer than the surrounding rural areas.
2. *The decrease in sunlight penetrating the air pollution layer.* The presence of, especially, particulates in the atmosphere prevents as much sun from reaching the surface on the earth. Not only can the particulates themselves shield some of our cities, but also, more importantly, the presence of more particulates tends to increase the cloud cover at higher altitudes, also preventing some sunlight from reaching the earth's surface.

Which of these two possibilities will predominate only time can tell, but an average change in either direction of only a few degrees could herald major climatic changes.

The magnitude of the temperature changes has been predicted to range from less than 0.05° to a maximum of perhaps 5°F. The temperature at any given location typically can vary this amount in less than an hour. Why then is there serious concern about these relatively small temperature changes that will neither freeze us to death nor produce unbearable heat?

The popular notion is that as the earth's temperature increases, the glaciers will melt, increasing the ocean levels 300 or more feet. This would be devastating, since most of the world's population lives at elevations of 1000 feet or less. For example, the highest point in Florida is only 180 feet above sea level. However, this melting would require centuries to be completed.

The most critical short-range impact would be shifts in the major crop-growing regions of the world. If average temperatures rise even 3°F, the major grain crops would have to be grown in more northerly regions, where the soil is much less productive. Violent weather associated with rapid climatic changes could also sharply reduce crop production. The world's ever-increasing population cannot tolerate any major reductions in grain production.

To date, there is no evidence for global warming or cooling. Extensive temperature data is being collected by the TIROS-N series of weather satellites.[12] The first decade's worth of information, from 1979–1988, has been analyzed and it revealed no long-term temperature trends. The satellite readings did show a lot of variability from month to month and year to year, refuting the popular perception that the earth's atmosphere is gradually warming on an annual basis.

Looking at longer-term terrestrial-temperature data, scientists did observe an increasing temperature trend prior to 1940, before the major rise in atmospheric CO_2 concentration. This was followed by a decrease in temperature from 1940 to 1965, when concern arose about a possible approaching ice age.[13] Clearly, more data is required before any firm conclusions as to long-term temperature trends can be drawn.

Another possible global effect (not truly a climatic effect, however) is related to the earth's O_3 layer. The presence of the O_3 layer is very important to life on our planet, for it helps shield us from potentially damaging ultraviolet (UV) radiation. A number of recent scientific studies have noted, particularly, a hole in the ozone layer located above Antarctica.

The depletion of the O_3 layer has been potentially associated with a number of pollutants. The nitrogen oxides can react with O_3 as such: $NO + O_3 \rightarrow O_2 + NO_2$ Normally, the nitrogen oxides do not reach the stratospheric O_3 for they react before diffusing that high, so this is not a major problem. NO_2 has a typical residence time of only three days, and NO, four days. However, the new supersonic transport planes (SSTs) emit nitrogen oxides directly into the stratosphere, right at the O_3 layer, so it is possible that the effects could be serious.

Chlorofluorocarbon (CFC) compounds, often used as propellants in aerosol cans and as refrigerants in air conditioners, can also have possible detrimental effects on the O_3 layer. Their relative nonreactivity, while making them good propellants, for example, allows them to migrate into the stratosphere before reacting. Many U.S. companies have voluntarily replaced CFCs by other chemicals or by mechanical pumping techniques, but a United Nations accord does not call for global elimination of CFCs until the year 2000, with an additional 10-year grace period for developing countries.

WATER POLLUTION

Water pollution often is a double problem for industries; frequently, it is necessary for the plant to condition the water before its use, as well as to treat the waste water after use. Pretreatment of the water is necessary to avoid a number of problems. The following is a partial list of some of the likely contaminants and the difficulties they are likely to cause.

Alkalinity: Typically biocarbonates, carbonates, and various caustics.
   Problems: Generates scale, particularly in boilers.
Hardness: Calcium and magnesium salts primarily. Other salts (for example, iron) are also minor contributors.
   Problems: Scale, particularly in boilers.
Sodium salts: Often as the sulfates, chlorides, nitrates, and bicarbonates.
   Problems: Very bad for particular industries, such as cellulose and drugs.
Silica (SiO_2):
   Problems: Scale.
Iron and manganese:
   Problems: Tend to stain. Very objectionable in the paper, textile, and tanning industries.
Aluminum:
   Problems: Usually not a problem for industrial purposes.
Fluorides:
   Problems: Usually not a problem, except, perhaps, in the production of baby foods.
Carbon dioxide:
   Problems: Increases the alkalinity; enhances the corrosive behavior of dissolved oxygen.
Oxygen:
   Problems: Corrosive to iron, steel, galvanized iron, and brass.
Nitrogen:
   Problems: Helps retain dissolved O_2 in water; thus increases corrosiveness.
Hydrogen sulfide:
   Problems: Rotten egg odor; corrosion of iron pipes, fittings, and equipment.
Methane:
   Problems: Fire and explosion hazard.
Microorganisms:
   Problems: Form coating in pipes; stains, tastes, and odors; decompose organic substances such as cellulose.
Organic matter:
   Problems: Causes tastes and odors, often absorbed by various processes; forms colored colloidal suspensions.

The discharge water from various industries has many of the same problems, plus some additional ones. The type of contaminant is dependent upon the particular industry and the particular process employed. We can classify the contaminants into three categories:

1. *Floating materials.* Typical floating materials would be oils and greases. They make the water unsightly; retard aquatic plant growth by blocking the sunlight and interfering with the natural reaeration; destroy the natural vegetation along the banks; are often toxic to fish and other aquatic life; destroy water fowl; and can also be a fire hazard.
2. *Suspended matter.* A common example of suspended matter is mineral tailings. Typically, mineral tailings form slime and sludge, which smother purifying microorganisms and ruin fish spawning and breeding grounds. If the suspended matter is organic, it would decompose using the dissolved oxygen and would produce noxious gases and odors.
3. *Dissolved impurities.* Typical dissolved substances would be acids, alkalies, heavy metals, and insecticides. In general, they make water undrinkable and destroy aquatic life. For example, phenols, even in very low concentrations (0.001 mg/L), give a very objectionable taste and odor. They also can build and concentrate their effects through the normal food chain. For example, an unpleasant taste is noticed when eating fish that had lived in water with a phenol concentration of only 0.0001 mg/L.

   Little is known about the effect on human health of many of the common water pollutants. Nitrates have been linked to methemoglobinemia and death in infants. The ingested nitrates can be reduced by microorganisms in the digestive tract to nitrites: $NO_3^- \xrightarrow{\text{microorganisms}} NO_2^-$. The nitrites can then oxidize the iron atom present in hemoglobin from Fe^{2+} to Fe^{3+}. The result is a methemoglobin molecule, one that is incapable of O_2 transport.

   Most of the other common substances probably have no major acute effects on human health. Many, however, do have chronic effects. For example, many of the organic substances are carcinogens. Selenuim causes bad teeth, gastrointestinal problems, and skin discoloration. Sodium and/or potassium are bad for people with certain health problems. On the other hand, some water impurities appear to be beneficial. A deficiency of chromium favors atherosclerotic diseases; there is evidence that degenerative cardiovascular disease is reduced by a factor present in hard water but absent in soft water.

PCB's: An Example of a "Common" Dissolved Toxic Pollutant

There are a number of particularly dangerous chemicals that have made their way into our water systems. Many of these chemicals are in industry effluents, or at least have direct industry sources.

Polychlorinated biphenyls (PCBs) are a well-known example of a toxic substance that has made its way into many of our natural water systems in this manner. PCB-contaminated rice caused an epidemic of a disfiguring skin disease and was blamed for 16 deaths and 2 stillbirths in Japan in 1968. As a result, the U.S. Food and Drug Administration (FDA) set maximum tolerable limits in food (for example, 5 ppm in fish).

PCBs were used for many years in "closed system" electrical applications, such as transformers. The manufacture of new, PCB-containing electrical devices is now illegal in the United States, and older design equipment is gradually being replaced. In spite of the controls now on PCBs, they are persisting in our environment. Other countries still use them in lubricants, duplicating paper, inks, paints and coatings, adhesives, plastics, and so forth, and much is being imported to the United States. The PCBs often leach from landfills, or get into the atmosphere by low-temperature incineration (PCBs do not burn until 2200–2600°F). In addition, accidents, often involving old transformers, appreciably contribute to the amount in our water systems.

The EPA requires high-temperature incineration of all PCB liquids drained from transformers and high- and low-voltage capacitors. The transformers, when drained of the PCBs, plus dredge spoils, municipal sewage sludge, and materials contaminated by spills, must also be incinerated or disposed of in a chemical landfill. Large capacitors could be placed in chemical landfills only until January 1, 1980.

The Lake Michigan trout have been declared unsafe because of their PCB content. It is illegal at the present time for commercial fishermen to catch trout in these waters. The FDA has power to seize any fish it believes to be contaminated. Advisory limits, based on fish number, size, and species, have also been established and widely advertised for sport fishing. The Canadians have also considered terminating commercial eel fishing in the St. Lawrence River because of the high PCB levels.

The exact health effects of PCBs are now being researched. Laboratory tests have led to liver cancer and reproductive failures in rats, chickens, minks, and rhesus monkeys. People who work with PCBs, including employees of many small manufacturers and transformer rebuild shops, have experienced nausea, skin disease, fungus infections, nose and eye irritation, and asthmatic bronchitis.[14]

Studies are ongoing to develop techniques for the removal of PCBs from spill sites. Some promising studies[15] have shown that it may be much simpler in some cases than people have thought. Evidence now being evaluated indicates that lime mixed with cement kiln dust (containing several forms of calcium and several trace metals) neutralizes the PCBs. The hope is that this mixture may be layered over the site, and within several days the material may be carted away for ordinary disposal. Scientists postulate that one or more of the trace metals catalyzes the degradation of the PCBs. If this method is validated, cleanup costs might be only 10–20% that of incineration.

Bioremediation, or decomposition using microbes, is another possible degradation technique that is now under study. This process is further discussed in Chapter 5.

BOD

A major problem associated with water pollution is the "biological oxygen demand," or the BOD. The BOD is the amount of oxygen required to biologically oxidize the water contaminants to carbon dioxide (CO_2), and thus is a measure of the suspended, colloidal, or dissolved organics. To measure the BOD level, the sample is usually allowed to incubate for five days. Within this period, about 70–80% of the actually present organic contaminants are oxidized by the microorganisms present. The choice of units used for BOD reflects the weight of O_2; hence, the units are typically pounds or kilograms.

The BOD is important in that the higher the BOD, the higher the organic content of the water, and the more dissolved O_2 that will be used to decompose these organics. A lack of dissolved O_2 in the waterways kills off desirable fish (for example, trout). Though there may still be plenty of fish, there usually is a marked change in the types of fish present if the BOD is high.

The presence of organic substances can decrease the dissolved oxygen in another way. The organics, along with the nitrogen and phosphorous, primarily can serve as food for algae (Figure 1-6). These algae are microscopic, greenish-colored plants that live in water. In themselves they are not harmful to humans (in fact, in limited amounts, they are useful, for they do create O_2 for other aquatic life by photosynthesis), but they do become aesthetically unsightly, a nuisance, and a hazard to other aquatic life when they proliferate. Their growth rate is dependent upon the conditions; they do not grow in deep, fast-moving, muddy, or cold waters, but flourish in lakes, ponds, and slow-moving streams as long as a sufficient supply of nutrients is available. Thus, the nutrients can act as the limiting factor in their growth. Over-fertilization with nutrients, or eutrophication, can occur naturally (for example, Green Bay, part of Lake Michigan, was totally covered by algae when first seen by white men), but man can accelerate the process.

The algae eventually die and decompose. In addition to producing a slimy scum and obnoxious odor, they require O_2 and hence can drastically decrease the dissolved O_2 supply as they decay.

The particular nutrient that is in limited supply acts as the "limiting factor" in algae growth and multiplication. Usually it is very difficult, if not impossible, to control the nitrogen. Farm runoff supplies much to the lakes and streams. Some algae are even nitrogen-fixing, and therefore can obtain all the nitrogen they wish directly from the atmosphere. Organics can be controlled somewhat, but usually the phosphorus is targeted as the nutrient to try to minimize (hence the emphasis on low-phosphorus detergents).

Manufacturing produces large quantities of BOD—20–25 million lb annually.

FIGURE 1-6.   Algae on a river downstream from a manufacturing plant. (*Photo courtesy Wisconsin DNR.*)

The major contributors are the chemical industry (about 44% of the manufacturing BOD), the pulp and paper industry (about 27%), and the food processing industries (about 20%). Table 1-2 lists typical BOD and suspended solids levels produced by a variety of industries.

Related to the BOD, and often measured instead, is the COD (chemical oxygen demand). The COD is a measure of the amount of oxygen required to *chemically* oxidize the contaminants to CO_2. The value of the COD is higher than the BOD because strong oxidants are used that force many substances to react that wouldn't react using biological microorganisms. The COD value thus represents almost 100% of the total organics present.

Dissolved sulfites and ferrous compounds, which can act as reducing agents, can also help deplete the O_2 supply.

Effect of BOD on Dissolved Oxygen Levels
in Streams

When a high-BOD pollutant is discharged into a stream, the level of dissolved oxygen (D.O.) downstream will decrease nonuniformly, but in a predictable

TABLE 1-2   Typical Process Waste Strengths[a]

Type of Industry	Process Wastes (mg/L)	
	BOD	Suspended Solids
Petroleum refining	310	370
Leather and leather products	1300	1700
Flat glass	310	370
Concrete, gypsum, and plastic products	310	10,000
Primary metal industries	310	370
Industrial launderers	870	1100
Meat products	800	520
Dairy products	2100	370
Preserved fruits and vegetables	1400	370
Grain mill products	3200	4500
Bakery, mixed	2500	1100
Fats and oils	1300	1100
Malt beverages	2100	980
Malt, wine, brandy, distilled liquor	650	370
Seafood and fish	1400	600
Food preparations	4100	930
Medicinals and botanicals, fermentation	4500	4700
Paints and allied products	2500	600
Industrial organic chemicals	310	370
Miscellaneous chemical products	2500	2600

[a]Courtesy Milwaukee Metropolitan Sewerage District.

manner. Soon after discharge, the organics begin to decompose and the D.O. level correspondingly drops. The rate of reaeration (the dissolving of oxygen from the air) increases, but generally not at a rate sufficient to prevent depletion of the oxygen in the stream. The resultant D.O. profile is illustrated in Figure 1-7.

The dip in D.O. is referred to as the "dissolved oxygen sag curve." If the concentration of BOD is sufficiently large (curve b), the D.O. level can temporarily drop to zero, and the stream become anaerobic.

At any location in the stream, at any time, there is an oxygen balance, which is determined by:

1. The amount of oxygen used by microorganisms for the decomposition of the pollutant; and
2. The reaeration rate.

The decomposition is usually assumed to be an exponential function influenced by, among other factors, the water temperature and the type of waste. The reaeration is represented by a complex function that involves the average stream velocity, the

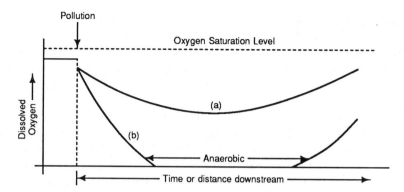

FIGURE 1-7.   Dissolved Oxygen Sag Curve. Curves (a) and (b) represent different amounts of BOD.

water temperature, and the oxygen deficit—that is, the difference between the current D.O. level and the saturation amount.

In general, the D.O. level is affected most in small ponds and sluggish streams; the effects of an equivalent pollutant in large, swift streams can be significantly less. Of most concern is the minimum D.O. level, which may occur several miles downstream from the pollution source. These minimum levels generally determine the type and numbers of aquatic organisms that can survive in the water system.

Acid Water Systems

Another problem often experience in natural water systems due to industrial pollutants is a change in the pH of the water. Natural waters now vary in pH from 4.0 to about 9.5. Some of the acidification of the waters occurs naturally. Carbonic acid, formed by the reaction of CO_2 and water in the atmosphere, is picked up in the normal precipitation process, so that rainwater is typically at about a pH of 5.6. However, acid mine seepage and, particularly, acid rain have had a major effect on the pH of many lakes and streams. Acid rain, for example, can have a pH as low as 3.9. In areas where calcium or lime are present in the rocks or soil, some of this additional acid is neutralized. In many areas, particularly in the high elevations of the Adirondacks, White Mountains, Green Mountains, Maine, and eastern Canada, this neutralization does not occur, and many of these lakes have a pH below 5.0. This problem is exacerbated by the fact that air pollutants generated in the industrial belt along the Great Lakes tend to fall in New England.

The pH of the water has a major effect on the type of fish and other aquatic species present. A recent study indicated that pH was 30% more important than water temperature and 50% more important than dissolved O_2, at least in the case of bass populations. Both trout and bass prefer alkaline streams, though that may be due somewhat to the additional advantage that more food is usually present. Alkaline streams are usually found in localities where the rock and soil contain $CaCO_3$. Rain, with its normally small amount of carbonic acid (H_2CO_3), dissolves the $CaCO_3$ present and forms calcium bicarbonate. The calcium bicarbonate is a source of CO_2 for photosynthesis by green plants. In addition, the $CaCO_3$ itself is necessary for the growth of crustaceans such as crayfish, scuds, and sow bugs, another source of food for fish. Trout can withstand somewhat acidic waters; in fact, pH values as low as about 5.0 are possible. However, in these conditions their growth is poor.

This effect on even fairly remote, high-elevation lakes is a serious problem of particular concern to the sports fisherman. Work is being undertaken by several groups to develop a means of neutralizing some of these acid lakes. The New York Department of Environmental Conservation has experimented with dumping agricultural crushed lime by helicopter into selected lakes, and has found it feasible on a cost/benefit basis. Studies are also underway to develop an acid-resistant strain of brook trout. The situation does not look as if it will improve, however, in the near future.[16]

Other Considerations

In general, industrial manufacturing is the leading source of controllable man-made water pollutants. In addition to the large quantities produced, industrial wastes are also much more likely than other types of waste to contain chemicals that are not biodegradable, toxic organic substances, and trace elements that must be removed.

When one considers water pollution control and management, it is necessary to consider, in addition to the nature of the contaminants, also how those substances travel through the water systems. As they travel, the contaminants may be degraded by chemical, physical, or biological processes. They migrate by flow or convection, disperse throughout the water systems by diffusion and/or mixing. In order to make the best decisions as to how to handle these waste waters, it is necessary to consider all of these processes and the overall transport. Only in this way can wise decisions be made as to which pollutants must be removed, and which could remain in the effluent. Computer programs have been developed for this purpose, and are taken advantage of by various industries.

The size of pollutant particles also affects their behavior. The sizes can be categorized as follows:

1. Settleable	$>100 \ \mu m$	largely organics
2. Supracolloidal	$1-100 \ \mu m$	
3. Colloidal	$1 \ nm-1 \ \mu m$	bacteria, viruses
4. Soluble	$<1 \ nm$	inorganics

The larger particles can absorb and bind various substances on their surfaces, and thus can act as (for example) sites to promote bacterial growth. Clays can absorb dissolved organics such as pesticides. These clays can then settle to the bottom of the lake or stream. In this way, the pesticides and other organics become incorporated into the sediments. The release of these pollutants back into the body of the water is unclear, and may pose a serious problem in the cleanup of our now polluted waters.

The intentional use of soil systems to strain and absorb wastes and to remove even bacteria and viruses has had some limited industrial applications. Often, the soil systems can degrade undesirable compounds to stable ions normally found in ground water—for instance, to nitrates, phosphates, carbonates, and sulfates. On the other hand, sometimes chromate ions, phenols, and many other "toxic" substances are not degraded, and can migrate underground many miles, contaminating water supplies far from the pollutant source.

SOLID WASTE

Solid waste disposal is, for industries, intimately connected to air and water pollution control. Typical solid wastes are mine tailings and cement dust. Considering these as examples, it is easy to understand the interrelatedness of these three types of pollution. Mine tailings (Figure 1-8), the remains of the mining operation

FIGURE 1-8.   Tailings produced by aluminum processing.

FIGURE 1-9.   Sludge generated by treatment of wastewater. (*Photo courtesy Wisconsin DNR.*)

after the mineral (for example, iron) is concentrated, are usually landfilled. The sulfates and other minerals that become exposed during mining react with rainwater, forming H_2SO_4 and thus becoming a water pollution problem. On the other hand, cement dust is generated during the cement manufacturing process, and is initially an air pollution problem. But collection of this dust turns into a solid waste disposal problem, and, if the dust is landfilled, it can leach alkalies into nearby natural water systems, thus becoming a water pollution control problem.

Other industrial solid wastes include fly ash, the unburned mineral matter from coal, coal dust, metallic foundry dusts, and sand from the foundry moldings and castings. Also included in the solid waste category are the sludges generated by treating polluted water (Figure 1-9). Many of these sludges contain toxic compounds that require special disposal techniques.

HAZARDOUS WASTES

Hazardous wastes are any waste that is toxic, flammable, reactive, or corrosive. Though generally liquids, they can also be solids, sludges, or gases. These wastes are generated primarily by the following types of industries, in order of descending volume: primary metals, organic chemicals, electroplating, inorganic chemicals, textiles, petroleum refining, and rubber and plastics.[17] Typically, the wastes are toxic metals (arsenic, chromium, lead, mercury, cadmium); salts, acids, or bases; or synthetic organics (halogenated hydrocarbon pesticides, PCBs, phenols).[18]

Hazardous wastes are currently perhaps the focus of the greatest concern by the public, government officials, and industry personnel. This is only reinforced because of the problems old (and some current) dump sites are now causing and the hazardous materials that are being released in this way to our environment.

References

1. "Nitrogenous Air Pollutants: Bigger Hazard." *Chemical and Engineering News*, March 27, 1978, p. 22.

2. "Can Air Pollutants Cause Chronic Lung Diseases?" *Environmental Science and Technology* **12**, *No. 13*, December 1978, p.1356.

3. McKee, Herbert C., "Particulate Standards Keyed to Visibility and Citizen Complaints are Interim Measure." *Environmental Science and Technology* **3**, *No. 6*, June 1969, p. 542.

4. "Inhaled Particulates." *Environmental Science and Technology* **12**, *No. 13*, December 1978, p.1353.

5. Hodgson, T. A., Jr., "Short Term Effects of Air Pollution on Mortality in New York City." *Environmental Science and Technology* **4**, *No. 7*, July 1970, p. 589.

6. ikens, Gene, "Acid Precipitation." *Chemical and Engineering News*, November 22, 1976, pp. 29-44.

7. "Acid Rain Network Planned." *Science News*, June 24, 1978, p. 407.

8. "Preserving Our Visibility Heritage." *Environmental Science and Technology* **13**, *No. 3*, March 1979, p. 267.

9. *Ibid.*

10. Jacobson and Hill (Eds.), *Recognition of Air Pollution Injury to Vegetation and Pictorial Atlas*. Pittsburgh: Air Pollution Control Association.

11. Ross, R. D., *Air Pollution and Industry*. New York: Van Nostrand Reinhold, 1972, p. 12.

12. "Method Outshines Data In Global Warning Study," *Chemecology* **19**, *No. 4*, May 1990, p. 6.

13. Singer, S. Fred, "What to Do About Greenhouse Warming." *ES & T* **24**, *No.8*, August 1990, pp. 1138-39.

14. *Great Lakes Focus on Water Quality* **2**, *Issue 1*, Winter 1976.

15. "Tests Find Lime Mix Neutralizes PCBs," *Insight* **7**, *No.15*, April 15, 1991, p. 49.

16. Rosenbauer, Tom, "The Meaning of pH in the Water You Fish." *Orvis News* **XIV**, *No.2*, March 1979, p.15.

17. Environmental Action Foundation, "Hazardous Wastes: Hidden Danger," *Garbage Guide* (RCRA Resource Guide) *No.2*, 1978, p. 2.

18. Environmental Action Foundation, "Caution: Timebombs." *No.6*, 1975, p. 1.

2

Regulations Governing Pollutant Emissions

The U.S. government's official concern about environmental quality was really initiated in 1970. Prior to this time, there were several regulations in existence, but there was no widespread coordination of efforts and little enforcement of the provisions. In January 1970, the National Environmental Policy Act was signed into law, and shortly thereafter the Council on Environmental Quality (CEQ) was established to coordinate the environmental activities in the United States. Then, in December of that year, the Environmental Protection Agency (EPA) was established.

The EPA has full jurisdiction to govern pollutants in the United States. It has the responsibility to do research, set standards, monitor emissions, and enforce the laws concerning pollutant emissions for all three categories: air, water, and solid waste. It gained its original authority through three laws: the 1970 Clean Air Act; the Solid Waste Disposal Act of 1965, and its amendment, the National Resource Recovery Act of 1970; and the 1899 Rivers and Harbors Act (Refuse Act), in conjunction with the Federal Water Pollution Control Act Amendment of 1972 and the Clean Water Act of 1977.

REGULATIONS CONCERNING WATER POLLUTION

The Refuse Act of 1899 prevented the dumping of wastes into the nation's navigable waters and gave enforcement responsibility to the U.S. Army Corps of Engineers. In 1972, President Nixon revived the Refuse Act—virtually ignored for over 70 years—with the 1972 Federal Water Pollution Control Act. At the core of the "Clean Water Act" is the Refuse Act Permit Program, which required some

40,000 industrial plants that discharged wastes into navigable waterways to specify the type and quantity of their effluents. In order to continue this waste water discharge, they had to apply for and obtain a waste discharge permit from the U.S. Army Corps of Engineers. All new industries must similarly obtain this permit. In evaluating the applications, the Corps of Engineers supposedly analyzes the impact of the proposed activity on the public interest, including water quality. A permit will be issued only when the benefits outweigh the foreseeable detriments.

To account for the size of various discharges, the Clean Water Act specified that each discharge should be considered on an individual basis. Minimum effluent standards were first set for all discharges, then these standards could be made more stringent based on the effect the discharge would have on the receiving water.

The goals established by this law originally called for the "best practical" control technology to be used in all industries by 1976 (as a general guideline, this technology must be widely used to qualify as the best practical), the "best available" by 1981 (even if the technique is not widely used by industry), and zero discharge by 1985. The scope of the regulations encompassed all waterways, municipal sewers, and deep wells. Oil drilling operations were exempt if the state had its own laws.[1]

As a result of this law, guidelines were set for many industries, listing the permissible amounts of various effluents. These guidelines were useful in considering the advisability of various discharge permit requests. Some of the industries for which guidelines are set include cement, lime, flat glass, gypsum, asbestos, leather tanning, beverages, grain milling, pulp and paper, canned and preserved fruits and vegetables, aluminum refining, and meat products. Various substances were restricted, depending upon the industry.

In addition, the Water Pollution Control Act allows individuals or organizations to sue the EPA, if they feel the EPA is not doing its duty in pollution enforcement.

The Clean Water Act of 1977 became effective December 27, 1977 and extended many of the provisions of the Water Pollution Control Act. The municipal waste water treatment construction grant program, previously in effect, was continued, but emphasis put on alternate waste water technology—recycling of waste water and sludge, land treatment, and methods to decrease waste water volumes. It also states "...it is the national policy that the discharge of toxic pollutants in toxic amounts be prohibited."

In June 1978, there was a court settlement between the EPA and several environmental groups. These groups—the Natural Resources Defense Council, Inc., the Environmental Defense Fund, Inc., the National Audubon Society, Inc., the Businessmen for the Public Interest, Inc., and the Citizens for a Better Environment—brought suit against the EPA for failing to implement portions of the Water Pollution Control Act. The settlement, commonly called the EPA Consent Decree, included the provision that the EPA publish a list of toxic pollutants for which emission guidelines and limitations be set. As a result, the EPA had to set effluent limits for about 100 toxic pollutants.

Certain priorities were required when attempting to determine effluent limits and test methods. Classed as "priority" pollutants were 129 chemicals, including all the toxic substances plus many of the common pollutants. These priority pollutants can be divided into nine groups: metals, asbestos, total cyanides, pesticides, total phenols, purgeable compounds, compounds extracted under acidic conditions, and neutral extractable compounds.[2]

To identify toxicants, the Clean Water Act of 1977 describes the National Pollution Discharge Elimination System (NPDES). NPDES allows both chemical and toxicity limits be set through permit programs. Initially, permits were based primarily on suspended solids and BOD; later, toxicity was monitored through chemical-specific analyses emphasizing the 129 priority pollutants. However, limiting the concentrations of priority pollutants in effluents did not ensure compliance with the Clean Water Act, since substances other than the priority pollutants can be toxic. Potential toxicants vary widely: polar and nonpolar organics, cationic metals, anionic inorganics, ammonia, and chlorine are all suspect.

In 1984, the EPA issued a policy statement recommending an integrated approach to toxicity identification. A whole-effluent toxicity test should be combined with chemical-specific analyses. For the whole-effluent test, the survival of aquatic organisms, such as various fish species or cladocerans, is determined.[3] This procedure is commonly referred to as "biomonitoring".

All industries have a choice of either treating their wastes and then discharging the treated waste waters to a natural waterway or of discharging their waste waters directly to a public sewer. In the latter case, the treatment problems are transferred to the publicly owned treatment facilities. Since some industrial discharges can cause severe treatment problems, tighter restrictions on what can be discharged into public sewers have been necessary.

The Consent Decree required new performance and pretreatment standards for 21 industrial categories, including timber products, steam electric power plants, leather tanning and finishing, iron and steel production, petroleum refining, organic and inorganic chemicals manufacture, nonferrous metals production, paint and ink production, pulp and paperboard mills, foundries, electroplating, ore and coal mining. For these industries, a corresponding list of specific contaminants and permissible effluent limits has been defined.

Local and/or state governments have set limits that apply more broadly, and control the effluent emissions of all industries, including non-listed industries, (for example, the canning industry). In general, these limits are less stringent than those identified by the federal regulations. The publicly owned treatment facility has the responsibility to enforce these regulations.

By the very early 1990s, the federal government started to phase out the municipal waste water treatment construction grant program, which had paid up to 75% of the cost of new facilities. Instead, the federal government provided seed money to the states, for them to establish a low case (2 1/2%) interest loan program.

Depending on current interest rates, this corresponds to a grant subsidy of approximately 25%.

REGULATIONS CONCERNING AIR POLLUTION

The 1970 Clean Air Act authorized the establishment of nationwide primary and secondary air standards, while allowing the previously established "air quality regions" to set stricter standards if desirable. Thus, through the establishment of performance standards, the Act limited the emissions on both old and new industries. The EPA may delegate to the states the enforcement of these standards, if the states' enforcement procedures are acceptable. Standards were established for SO_2, CO, NO_x, hydrocarbons, photochemical oxidants, and a number of other hazardous substances.

In addition to providing authority for establishing emission standards for industries, the Clean Air Act provided for many other actions. The most publicized is probably the demand for 90% auto emission reduction by, originally, 1975 (this deadline has been several times extended). The Act also: 1) strengthened enforcement procedures (the EPA can go to court to enforce the emission standards, and fines up to $25,000 or one year in prison are possible for each intentional violation); 2) allowed the EPA to take over and operate what is considered an unsatisfactory state plan upon giving 30 days notice; 3) provided for citizens' suits against various companies; 4) required the "best available" control technology on all new sources, compliance being enforced, of course, only after the plant is operating; 5) allowed for new developments in control technology developed by one firm to be ordered to be made available to other companies if there would otherwise be a violation of the standards by the second company; and 6) provided for citizens' suits against the regulatory bodies, to sue the administrator to enforce the regulations except in cases where the administrator has discretionary power.

In 1977, Congress amended the Clean Air Act, giving the EPA the power to enforce federal air pollution standards and the responsibility of attaining the ambient air quality standards.

A second revision of the Clean Air Act became law in November 1990, after many years of Congressional debate. Among the other provisions, the Act requires:[4]

1. An increase, from 7 to 189, of the number of toxic chemicals to be controlled, with a goal of 90% reduction by the year 2000;
2. A 15% reduction in ground level ozone in polluted areas by 1996;
3. Phasing out of chlorofluorocarbons and carbon tetrachloride by 2000 and banning hydrochlorofluorocarbons by 2030;
4. Lowering SO_2 and NO_x emissions from fossil fuel burning to reduce acid rain; and

5. Establishing an independent chemical safety board to investigate chemical accidents.

To achieve the planned reduction in emissions, all major sources (plants emitting . 10 tons/year of any listed pollutant or a 25 ton/year total of all listed pollutants) must install the best available pollution control devices, called Maximum Achievable Control Technologies or MACT. The law will not apply immediately to all industries: various types of plants are being phased in during the period 1990–2000.

REGULATIONS CONCERNING SOLID WASTE

The Solid Waste Disposal Act of 1965 established the Office of Solid Waste Management, now operating under the EPA. It is responsible for:

1. Initiating and/or accelerating a research and development program emphasizing recovery and reuse; and
2. Technically and financially aiding states, communities, and private agencies in developing programs for storage, transportation, processing, recycling, and otherwise disposing of solid wastes.[5]

More significant for the environment, perhaps, are the solid waste regulations developed under the Resource Conservation and Recovery Act (RCRA). This statute, though popularly associated with hazardous waste disposal, also addresses nonhazardous solid waste. Several concerns of the public (and Congress) were addressed:[6]

1. Protection of public health and the environment from solid waste disposal;
2. Filling loopholes in existing air and water quality laws;
3. Providing proper land disposal of residues from air pollution and water pollution cleanup; and
4. Promoting resource recovery.

Disposal sites are classed by RCRA as either landfills, lagoons, or landspreading operations. The potential adverse effects of improper disposal includes floodplains, groundwater, surface water, endangered species, food chain crops, air quality, and human health and safety.

Floodplains—Disposal facilities are restricted from floodplains unless the area has been protected against being washed out by a "100-year flood."
Groundwater—Owners and operators of disposal facilities must comply with

proper design requirements, such as collection and proper disposal of leachate, utilization of correct cover material, and adequate groundwater monitoring.

Surface water—Proper facility design and maintenance are required, often including liners, levees, and dikes.

Endangered species—The disposal facility must not contribute to the hunting, trapping, capturing, or otherwise harming of endangered or threatened species.

Food chain crops—The operator of a landspreading operation must not permit the movement of heavy metals and synthetic organics (such as polychlorinated biphenyls (PCBs)) into soils used for human food chain crops. This can be controlled by limiting the solid waste dosage or by suitable selection of crops.

Air quality—A disposal facility must comply with the Clean Air Act and relevant state and local laws. Basically, this outlaws open burning except for selected debris.

Health and safety—Generally, two dangers are addressed: fires and the danger of birds interfering with aircraft. Fire danger is largely eliminated by the ban on open burning; aircraft safety can be assured by proper disposal site location.

REGULATIONS CONCERNING HAZARDOUS WASTES

RCRA

Hazardous wastes can be considered to constitute a category unto themselves. They are usually liquids, but can also be solids, sludges, or gases. The Resource Conservation and Recovery Act (RCRA), implemented in the fall of 1976, allows the EPA to regulate these wastes "from cradle to grave."

The EPA really had three tasks to accomplish in accordance with this act:

1. It defined what characteristics make a waste hazardous (toxicity, flammability, etc.) and listed which specific wastes are to be governed as hazardous wastes.
2. It developed standards that addressed all aspects of handling, transporting, and disposing of hazardous wastes.
3. It could additionally authorize state-run programs for either partial or full enforcement of the hazardous waste regulations if such programs are at least equivalent to the federal program.

In November 1984, the Hazardous and Solid Waste Amendments (HSWA) to RCRA became law.[7] These amendments required the EPA to evaluate all "listed" and "characteristic" hazardous wastes according to a strict schedule, to determine which wastes should be restricted from land disposal. For the restricted wastes, the EPA had to set levels and/or methods of treatment that substantially diminish their toxicity or reduce the likelihood that hazardous constituents from the wastes will migrate from the disposal site. These requirements took effect for the last group of such wastes in May 1990.

For the purpose of these amendments, land disposal includes placement in:

- Landfills
- Surface impoundments
- Waste piles
- Injection wells
- Land treatment facilities
- Salt domes or salt bed formations
- Underground mines or caves.
- Concrete vaults or bunkers, intended for disposal.

Different treatment standards have been established for two separate types of solvent wastes: wastewaters (solvent-water mixtures containing less than or equal to 1% w/w total organic carbon) and all other spent solvent wastes, including solvent-containing solids and soils.

The wastes are classified according to source. For example, F007 wastes are spent cyanide plating bath solutions from electroplating operations; K061 wastes are emission control dusts/sludges from the primary production of steel in electric furnaces; and K083 wastes are distillation bottoms from aniline production. A given chemical—(for example, pentachlorophenol (PCP)) might legally be land disposed in quite different concentrations, depending on the source of the waste.

The allowable emission standards are based on the demonstrated performance of treatment technologies, such as incineration, steam stripping, biological treatment, and activated carbon treatment. The specific emissions standards are expressed in terms of extract concentrations obtained by employing the Toxicity Characteristic Leaching Procedure (TCLP). This is an analytical method adopted to determine whether the concentrations of hazardous constituents in the waste extract or an extract of the treatment residual meet the treatment standards.

The EPA does not require that a specific technology be used to meet the established treatment standards, only that the wastes meet the concentration standards prior to land disposal. Dilution is prohibited as a substitute for adequate treatment; however, it is permitted as a necessary part of a waste treatment process.

Superfund

In order to address the large number of hazardous waste sites in the U.S. that were emitting hazardous materials, in 1980 Congress enacted the "Comprehensive Environmental Response, Compensation, and Liability Act" (CERCLA), more commonly known as "Superfund." With this act, the EPA was given a broad mandate to cleanup hazardous waste sites, as well as $1.6 billion to do the job.

In 1986, the program was extended for another 5 years under the auspices of the Superfund Amendments and Reauthorization Act (SARA), and an additional $8.5 billion was authorized. This money was to be raised by a general manufacturing tax, specific levies on chemical and petroleum products, and from general treasury funds, and was to be used to pay for cleaning up abandoned hazardous waste sites for which the responsible parties could not be found and assessed the costs.

In addition, a new "Leaking Underground Storage Tank Trust Fund" was established by levying a new 0.1c/gal tax on all motor fuels. Community emergency planning and right-to-know provisions were also initiated.[8]

As a results of these 1986 amendments, the EPA also established a proposed cleanup schedule, citing 175 new remedial action starts to be initiated by October 1989 and an additional 200 by October 1991.[9] By 1989, it had identified some 1175 sites that it placed on a "National Priorities List" (NPL).

Severe criticism has been levied against the EPA in the handling of this cleanup. This criticism has included accusations of a lack of priorities, a lack of internal agency coordination, an inability to select proper cleanup procedures, delayed cost recovery actions, and overall tremendous costs. Cleanup costs for the current NPL sites alone are estimated to be $30 billion; costs for additional sites that will probably be added to the NPL (which may eventually total 4000) are estimated at another $55–74 billion.

A study by the Oak Ridge National Laboratory Office of Risk Analysis found, in addition, that 70% of all the Superfund NPL sites prior to cleanup had risk levels actually less than the target risks that were established for the sites after remediation.[10]

Hazardous waste sites are not going to miraculously disappear. Hopefully, experience and education will both permit proper cleanup of the existing sites and prevent the occurrence of the next generation of problem locations.

Underground Storage Tanks (USTs)

In 1989, the EPA established rules to protect groundwater from leaking underground storage tanks. Though the EPA estimated that half a million tanks were leaking, little was done to enforce the rules until multi-million gallon leaks were discovered during the summer of 1990 in Brooklyn, NY and Hartford, IL.[11] Those incidents triggered much concern and attention about the problem.

In accordance with EPA rules, all tanks must be equipped with leak detectors after 1993; by 1998, all hazardous chemicals must be stored in double-walled tanks, and the space between walls must be monitored. To prevent corrosion, new tanks must be made of either coated and cathodically protected steel, fiberglass, or steel clad with fiberglass. To upgrade existing tanks and pipes, cathodic protection or an interior lining, or both, can be added. The UST rules also specify proper procedures for installing and closing tanks, and for reporting leaks.

Compliance costs are not insignificant. Westinghouse Electric Corp. paid $150,000 to simply remove 12 steel tanks from its resin-coating plant in Bedford, PA; Eli Lilly and Co. paid $140 million to replace 272 tanks used to store solvents and intermediates at two of its laboratories. Alternatives to USTs that companies are considering include above-ground storage tanks and monthly delivery of chemicals in drums or recyclable bulk-packing totes.

POLLUTION PREVENTION ACT OF 1990

In 1988, the EPA set up an Office of Pollution Prevention within the Office of Policy, Planning and Evaluation. This action was formalized when Congress passed the Pollution Prevention Act of 1990. Shortly thereafter, the EPA issued its strategy for pollution prevention activities. Key to this scheme is the Industrial Toxics Project, or the 33–50 Program.

In February 1991, the EPA asked more than 600 U.S. companies to voluntarily reduce pollution caused by 17 toxic chemicals: benzene, cadmium, carbon tetrachloride, chloroform, chromium, cyanide, dichloromethane, lead, mercury, methyl ethyl ketone, methyl isobutyl ketone, nickel, tetrachloroethylene, toluene, trichloroethane, trichloroethylene, and xylene. All are high-volume, high-release industrial chemicals. In 1988, these 17 chemicals accounted for 1.4 billion lb of the toxic wastes emitted from some 11,000 U.S. plants. Greater than 70% were emitted as air toxics.

In accordance with this program, the goal was to reduce the emissions of these chemicals 33% by the end of 1991 and 50% by the end of 1995, when compared to 1988 emissions.

The emphasis of this effort is in this order: source reduction, in-process recycling, treatment, if necessary, with disposal as a last resort. According to EPA guidelines, source reduction can be done by material substitution, product reformulation, process modification, and improved housekeeping.

This program, though voluntary, has appealed to many companies, particularly chemical companies, who are the greatest source of toxic substances. There is a definite cost advantage to compliance: less waste management, reduced use of raw materials, minimized liability, less regulatory burden, and even positive publicity regarding their efforts.

ESTABLISHING EMISSION STANDARDS

The establishing emission standards is always a very difficult job. Among the items that must be considered are the following:

1. What weight should be put on projected health effects based on data available from only the lower animals, typically mice and rats, and sometimes also

monkeys? Often, these animals react very differently to a substance than would a human; for example, some can metabolize a carcinogen to a harmless substance, something a human may not be able to do. Humans are not used for this type of testing, and epidemiological studies are scarce.

2. What fraction of the population should be protected? Many pollutants much more seriously affect the very young, the old, or those with serious illnesses such as emphysema. The danger levels are very different depending upon what percentage of the population we consider.

3. What is the effect of more than one pollutant? The combined effects of two or more pollutants—the synergistic effects—are often much more devastating than the sum of the individual effects.

4. Should agricultural interests be strongly preserved? It may be that the allowable pollutant levels must be more restricted for agricultural purposes than for human health.

5. What weight would be given to primarily psychological factors such as visibility?

6. What would be the economic and/or social impact that would result from a particular restriction? Would it be so economically unfeasible that some plants would have to shut down, putting hundreds out of work? This has happened in several cases in recent years. Tough antipollution laws are very desirable for communities, yet most cannot afford to discourage existing or potential industries. Increased pollution cleanup has led to increased production costs and hence to higher costs for some products, also.

Some industries are, fortunately, finding that they can save money by making their processes more efficient, and are simultaneously decreasing their emissions. The degree to which the emissions must be reduced is often a major factor in this; as the amount of permissible emissions decreases, the cost to remove those last few percent rises astronomically. Table 2-1, calculated by Sun Oil Co., based on data from the CEQ and the EPA, indicates the great rise when considering water pollution. These costs are only for comparison purposes; actual current costs are significantly higher.

TABLE 2-1   Costs for Pollution Reduction[a]

% Reduction in Water Pollution	Total Cost ($ Billions)	Cost per Incremental % ($ Billions)
85-90%	$ 61	$ 0.70
95-99%	119	6.00
100	317	66.00

[a]Courtesy the American Chemical Society, from the June 19, 1972 issue of *Chemical and Engineering News*.

Similarly, it is often more costly to install in existing plants equipment to remove the last remaining particulates, or other pollutants, than it would be to remove the same from new installations. The question that we must ask is if the cost to remove that few remaining percent bears a reasonable relationship to the benefits received.

An added difficulty in cleaning up our environment is the relationship with energy usage. At least on a short-term basis, energy conservation and pollution control are often contradictory policies. For example, during the nationwide coal strike during the winter of 1978, states such as Ohio were put on emergency status, and their industries were allowed to exceed the air pollution emission levels because of a drastic coal shortage. It was expected in 1970 that a 15% increase in the use of natural gas would allow industries to meet the particulate and SO_2 emission standards, in particular. This, of course, was not possible; in fact, many industries once on natural gas had to switch to coal, with the associated increase in emissions. Even the development of new energy sources required an environmental penalty.

The relationship between pollution control and the energy situation also has another facet. One of the major techniques of pollution control in many industries is to switch from one type of process to another type, one that generates fewer pollutants, or at least pollutants that can be more easily or economically controlled. Sometimes this conversion to a different process reduces the energy consumption, such as the conversion from wet to dry cement processes. In other situations, the energy requirements are essentially the same, as experienced in the conversion between sulfite and kraft paper pulping processes. In some industries, however, the results are not so promising. Organic solvents, such as those used in the printing and industrial coatings industries, do create serious air pollutant emission problems. But the use of waterborne coatings and inks, considering the overall application process, usually consumes significantly more energy than does the use of solvent-based materials. It is hoped that most changes in the type of process for many industries can, in the future, combine energy savings and pollution control.

There are several other regulations that also have an impact, albeit more indirect, on industrial pollution control. For example, it is feared that the Toxic Substance Control Act, also administered by the EPA, will slow down the pace of research and development of new materials to replace high-emission substances. Any chemical company that wishes to place a new raw material on the market must determine that the product is nontoxic before it can be introduced. This costly evaluation process will not be undertaken unless the company is assured of an adequate market for the raw material, since the toxicity tests typically cost more than one-quarter million dollars.

Though the EPA is the major federal agency with which an industry must deal when considering emission problems, there are several other federal agencies that also have an impact on industrial pollution control. For example, the Occupational Safety and Health Administration (OSHA) plays a key role when an industry attempts to switch processes to lower the emissions. This is a particularly serious consideration for industries that employ the use of organic chemicals and/or heavy metals such as does the chemical coatings industry.

In addition to the federal agencies, each state has its own agencies and sets of regulations. These regulations, in order to have any validity, must be at least as strict as the federal guidelines.

In past years, there had been a certain lack of government intercooperation with respect to some of these regulations. There had been cases where a particular company was working with the federal government in trying to clean up its wastes, when suddenly the local government turned around and slapped the company with a fine. Another potential difficulty was the somewhat indiscriminate use of penalties of all types levied against businesses. As an example, one can consider plant safety rules such as on noise control. Sometimes the penalties had been handed out with little reason, reflected in that two-thirds were later dismissed or reduced. This indiscriminate fining in other pollution control areas could have very serious consequences for not only that particular business, but for the total economy. The overall effect, to date, of various pollution regulations will be considered in Chapter 17.

References
1. Section 404, Permit Program, U.S. Army Corps of Engineers, September 1975.
2. Keith, Larry, and Telliard, William, "Priority Pollutants." *Environmental Science and Technology* **13**, *No. 4*, April 1979, p. 416.
3. Burkhard, Lawrence P., and Ankley, Gerald T., "Identifying Toxicants: NETAC's Toxicity-Based Approach." *Environmental Science and Technology* **23**, *No. 12.* Dec 1989, pp. 1438-43.
4. Zahodiakin, Phil, "Puzzling Out the New Clean Air Act," *Chemical Engineering*, Dec 1990, **97**, *No. 12*, pp. 24-27.
5. "Bureau Attacks Nation's Solid Waste." *Environmental Science and Technology* **3**, *No. 8*, August 1969, p. 705.
6. Vesilind, P. Aarne, and Peirce J. Jeffrey, *Environmental Engineering*. Boston: Butterworth Publishers, 1982, p. 388.
7. Land Disposal Restrictions: Summary of Requirements, U.S. EPA, June 1989.
8. Dowd, Richard M., "Finally... Superfund." *Environmental Science and Technology* **20**, *No. 12*, December 1986, p. 1207.
9. Miller, Stanton S., "Superfund: An Environmental Boondoggle." *Environmental Science and Technology* **23**, *No. 4*, April 1989, p. 394.
10. Travis, Curtis C., and Doty, Carolyn B, " Superfund: A Program Without Priorities." *Environmental Science and Technology* **23**, *No. 11*, November 1989, pp. 1333-34.
11. Shelley, Suzanne, "Out of Sight—Not Out of Mind." *Chemical Engineering* **98**, *No. 1*, January 1991, pp. 30-35.
12. Ember, Lois R., "Strategies for Reducing Pollution at the Source are gaining ground." *Chemical and Engineering News* **69**, *No. 27*, July 8, 1991, pp. 7-16.

3

Methods for Air Pollution Control

When one considers air pollution control, there are various approaches that can be taken. The potential pollutant can be controlled at the source, by one technique or another; it is also sometimes possible to dilute the pollutant so that its concentration would no longer be harmful. Dilution can be seen to be feasible when one remembers that for most pollutants the anthropogenic sources are only a small percentage, say 10%, of the total present in our atmosphere.

Considering first the controlling of the pollutant at the source, the most effective method, if possible to implement it, is to never generate the pollutant at all, or at least to minimize the amount. This can sometimes be accomplished by changes in raw materials, operating conditions, type of equipment, or even the drastic step of totally changing the process. This type of action can often be more economical overall than installing a large amount of very expensive equipment to remove the pollutants that have been formed.[1] If the pollutant cannot be prevented from forming, equipment that destroys, masks, counteracts, or traps the pollutant is required. Some pollutants can be destroyed by combustion or by catalytic action. Masking of odors merely makes a bad-smelling pollutant undetectable by superimposing another—preferably pleasant—odor. Counteracting of the pollutant also works for odor: two odors mix, cancelling each other out. Neither method removes the pollutant itself. Collection of the pollutants before they get into the atmosphere is the most commonly used method for reducing the air emissions that do form.

As could be expected, the particular type of air pollution control equipment used depends upon the nature of the pollutants: their size, shape, density, stickiness, and electrical properties. But it depends upon other conditions, as well, such as temperature, moisture content, quantity, and various economic factors. Most of the collection equipment is designed to handle particulates; some equipment can simultaneously also treat other types of pollutants; and some treatment methods are best for nonparticulate emissions. There are several general types of equipment that are used to handle the variety of emissions that exist. There are those that collect particulates: 1) dry mechanical collectors, 2) fabric collectors, 3) wet scrubbers, and 4) electrostatic precipitators; there are fume incinerators, which destroy the pollutants; there are sorption techniques, which either adsorb or absorb, particularly, volatile pollutants; and there are tall stacks, which act as a dilution technique. Among each general type there are usually many variations.

PHYSICAL PRINCIPLES

In terms of collecting particulates, there are several basic physical principles that can be, and are, employed in the operation of the various types of collectors:

1. *Gravity settling*. Small particles are carried along in a moving air stream. When the velocity of that air stream is reduced, many of the particles, particularly the heavier ones, can settle to the bottom of the collector.
2. *Inertial forces*. As the direction of a moving air stream changes, the heavier particles have a tendency, due to their momentum, to continue in a straight line. As they leave the air stream, they typically collide with a wall and then settle to the bottom of the collector.
3. *Filtration*. The air stream, saturated with particulate matter, passes through a porous material. The particulates are retained on the surface, and the now clean air passes through.
4. *Electrostatic attraction*. Particles can be electrostatically charged. These charged particles are then attracted to objects of the opposite charge, and removed from the air stream. When the two objects make physical contact, the particles are neutralized and fall to the bottom of the collector.
5. *Particle enlarging*. When dust particles pass through a water spray, a particulate—a water droplet agglomerate—is formed. The heavier weight of this agglomerate allows it to be more easily separated from the air stream. A very large quantity of water can virtually wash the particulates from the air stream.

All of the collection methods employ one or more of these principles in their operation. The efficiency of the collection tends to correspond to the size of particle: the larger, heavier particles are much more efficiently removed by all of these techniques.

DRY SYSTEMS

Efficiency

To compare various pollutant-removal systems, it is useful to look more closely at the concept of efficiency.

Consider the following sketch:

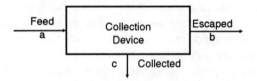

After the pollutant enters a (generic) collection device, there are two options: it can be collected, or it can escape from the device, generally to be emitted to the atmosphere.

A simple mass balance gives us:

$$Ma = Mb + Mc$$

where Ma = amount of pollutant into collection device
      Mb = amount of pollutant not collected
      Mc = amount of pollutant collected,
with all measured in kg/s or equivalent units.

The efficiency, η, of the collection device is the percentage of the total pollutant that is collected, and thus is correspondingly given by

$$\eta = Mc/Ma \ (100).$$

But, since Ma = Mb + Mc, this can also be written as

$$\eta = Mc \ (100)/(Mb + Mc).$$

In some circumstances one might know, instead of the mass flow rates, the air flow rates, Q (m³/s), and/or the concentration of the pollutant(s), C (kg/m³). Since $M = Q \cdot C$, given this information, the collection efficiency can be calculated by either

$$\eta = QcCc(100)/(QaCa)$$

or

$$\eta = Cc(Ca-Cb) \cdot 100/[Ca(Cc-Cb)].$$

Which one of these forms is used would be determined by the information available; the results are identical.

The various collection devices vary in their collection efficiency and the nature of the pollutants. In general, the simpler, less energy-intensive devices are less efficient, making them suitable mainly for large, dense particulates. Finer particulates and other types of air pollutants require the more sophisticated collection devices.

Dry Mechanical Collectors

Dry mechanical collectors provide a feasible technique only if one is dealing with particulates, and if there is no more than a light to moderate amount of them. Preferentially, the particles are fairly large. Often, a dry mechanical collector is put ahead of a more efficient collector, in order to remove the bulk of the heavier particles and thus to increase the overall efficiency and reduce the maintenance of the system.

These types of collectors require minimal power consumption and experience no corrosion unless a corrosive mist is incident upon them, but hot and hydroscopic emissions can present problems. In addition, the equipment is often large and very bulky.

Gravity Settling Chambers

Gravity settling chambers (Figure 3-1) are the oldest and simplest type of particulate collector. Particles between 40 and 100 μm in diameter are readily collected

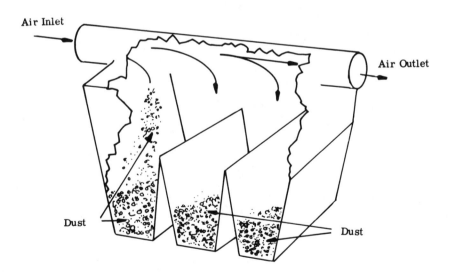

FIGURE 3-1.   Gravity settling chamber. (Adapted from material appearing in the *1972 Report on Business and the Environment,* edited by Price, Ross, and Davidson, a McGraw-Hill publication.)

by this technique. Their efficiency is very poor on fine dust (dust of less than 10μm diameter) and decreases as the load increases. They have a very large physical size, but, on the other hand, they are inexpensive and reliable, and they cost little to maintain and operate.

Recirculating Baffle Collectors

The addition of baffles or deflectors to a gravity settling chamber (Figure 3-2) increases the collection efficiency. The incoming gas must make rather an abrupt turn in traveling to the outlet, and this change in direction allows the larger dust particles, because of their momentum, to escape the air stream and more readily settle to the bottom of the collector. The physical size of baffle collectors is smaller than that of plain gravity settling chambers, and they are more efficient, even at higher loads. They can be used, in general, to readily collect particles between 10 and 50 μm in diameter. As with the settling chambers, their installation and maintenance costs are low, and they have high reliability. Unfortunately, in spite of their increased collection efficiency, they are not sufficiently effective to act as the only collector for an industrial process. In order to meet the emission control standards, a more efficient collector must always be installed downstream, hence allowing the baffle collector to serve as a "precleaner."

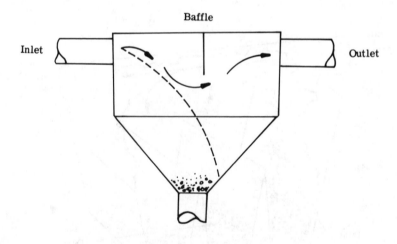

BAFFLE CHAMBER

FIGURE 3-2.   Gravity settling chamber with added baffle. (From *Pollution Control Technology.* New York: Research & Education Association.)

Cyclones

Cyclones (Figures 3-3 and 3-4) operate on the principle of centrifugal force. As the carrier gas enters the cylindrically shaped collector, it takes a helical path and the inertia of the particles carries them to the walls from where they drop into a hopper at the bottom. The clean gas rises near the center and is exhausted through the top. The cyclone collector is quite simple and reliable, with low initial cost, easy maintenance, and high temperature capabilities. Cyclone collectors tend not to be efficient on particles smaller than 10 μm, and are best for 15–50 μm particles. In contrast to gravity settling chambers, they are more efficient with heavier loads due to the increased interparticle interactions.

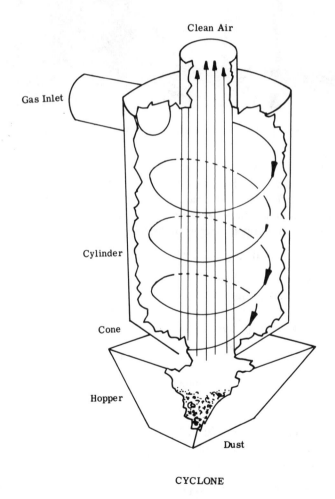

CYCLONE

FIGURE 3-3.   Cyclone collector.

FIGURE 3-4.   Cyclones used for particulate collection. (*Photo courtesy Wisconsin DNR.*)

The efficiency of a cyclone can be estimated by using the concept of "cut diameter," d_c, which is defined as the particle diameter at which 50% of the particles are removed. A graph, such as indicated in Figure 3-5, can be used to determine the collection efficiency for any size particle, assuming the cut diameter is known.

The cut diameter is defined in terms of the collection parameters by

$$d_c = [9b \, \mu/2\pi \, Nv_i \, [\rho \, s - \rho])]^{1/2}$$

where b = cyclone inlet width, m

μ = gas viscosity, kg/m · hr

N = effective number of turns of the particle in cyclone (~4)

v_i = inlet gas velocity, m/s

ρ_s = particle density, kg/m³

ρ = gas density, kg/m³.

A more accurate estimate of the number of turns (N) also can be calculated

$$N = \pi \, (2L_1 + L_2)/H$$

where H = height of the cyclone inlet, m

L_1 = length of cylinder section, m

L_2 = length of cone section, m.

If, then, the appropriate collection parameters and the average size particle are known, the approximate collection efficiency can be calculated.

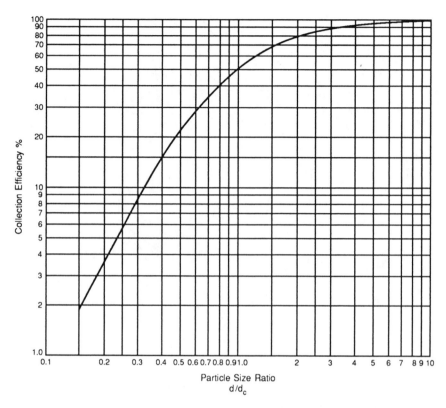

FIGURE 3-5.   Cyclone Efficiency. The ratio of particle size diameter (d) to the "cut diameter" or that particle diameter at which 50% of the particles are removed can be used to determine the removal efficiency of any size particle.

Multiple Cyclones

Since the centrifugal force on the particulates is given by

$$f_{cent} = \frac{mv^2}{r},$$

where m = mass of the particles

v = velocity of the particle

r = radius of path (approximately that of the cyclone),

it can be seen that the collection efficiency would increase with a decrease in the collector radius. A "bank" of cyclones (Figure 3-6), parallel with the feed, can thus be used more effectively than one single, larger-diameter cyclone. These smaller cyclones, of course, tend to plug more readily than those of larger diameter. Smaller particles can be collected with multiple cyclones, typically particles 5–20 μm in diameter.

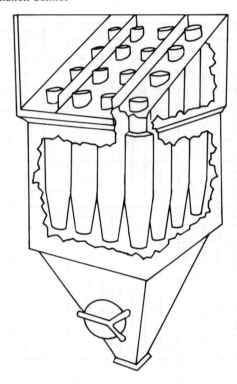

MULTIPLE CYCLONE

FIGURE 3-6.   Multiple cyclone collector. (From *Pollution Control Technology*. New York: Research & Education Association.)

Other Cyclone Modifications

The addition of louvers or vanes along the axis of a cyclone collector can increase the efficiency, particularly for coarse particles. The closer the spacing, the higher the efficiency. These added vanes do, however, suffer much abrasion and corrosion.

A fines "eductor," a small outlet that starts at the top of the cyclone and spirals downward around the outside, can be added to remove some of the very fine particles that are carried upward in the supposedly clean air being emitted. This addition increases the collection efficiency without increasing the operating cost.

Dynamic Precipitators

Dynamic precipitators (Figure 3-7) also operate on the principle of centrifugal force. They are really a combination of a specially designed centrifugal fan and a dust collector. The centrifugal force generated by the rotating blades pushes the particles in the air stream to the tips of the blades, from where they are drawn off in a concentrated stream. This type of collector is unsuitable for sticky or fibrous

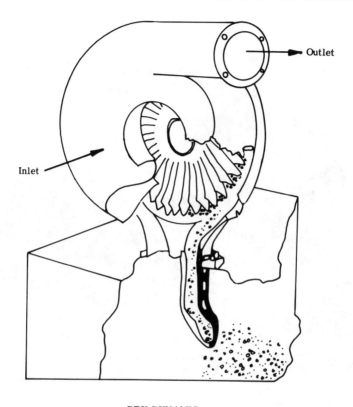

DRY DYNAMIC

FIGURE 3-7.   Dynamic precipitator. (From *Pollution Control Technology*. New York: Research & Education Association.)

materials, because the solids tend to build up on the blades. It is fairly efficient, however, even on particles 5–20 μm in diameter.

Fabric Collectors

Fabric collectors (Figure 3-8) provide a very efficient method for collecting particulates, even those particles less than 0.5 μm in diameter. All types operate on the principle of trapping the dust on one side of the fabric, the dirty gas side, while allowing the air itself to pass through the interstices in the fabric. As the dust collects on the side of the fabric, the collector becomes more efficient, because the dust layer—the filter cake—is even more efficient as a collector than is the cloth. This filter cake blocks the larger of the fabric interstices, which are typically about 100 μm in diameter, and thus helps to capture the finer dust particles.

FIGURE 3-8.   Baghouse at Airco's ferroalloy plant in Kentucky. (*Photo courtesy Airco, Inc.*)

The fabric must be cleaned frequently, or eventually no gas would be able to pass through it. The particular cleaning method chosen is a basis for classifying the types of fabric collectors. The three possible cleaning methods are:

1. Intermittent
2. Periodic
3. Continuous.

Intermittent cleaning requires that the collector be totally shut down every so often for removal of the dust cake from the fabric bags or envelope. During this shutdown period, the particulates are, of course, not collected. The frequency of the cleaning required would depend upon the dust load and the type of fabric used.

Periodic types of fabric collectors are each really a series of intermittent types, connected by inlet and outlet manifolds. Typically, they would have three sections; alternate sections would shut down every 30 minutes for a 2-minute cleaning. This shutdown normally is cycled automatically.

Continuous types are more expensive initially, but they are capable of handling heavier dust loads. Felt fabric is usually employed for this type of collector because the continuous cleaning prevents any significant dust cake development. Felts are more efficient than are the woven fabrics in collecting the very fine particles.

There are various methods of fabric cleaning available. These possibilities include the following:

1. *Manual or powered shaking.* (Figure 3-9 and 3-10.) The fabric bag is suspended on a flexible support. When it is desired that the filter cake be removed from the interior surface of the bag, that bag can simply be shaken. The attached dust cake breaks apart and falls into a hopper at the bottom.
2. *Air jet.* (Figures 3-11 and 3-12.) A high-velocity stream of air is blown into the collector bag opposite the normal direction of air flow. This dislodges the particles on the outside surface, which then fall into the dust hopper.
3. *Blow ring.* (Figure 3-13.) A traveling blow ring consists of a narrow hollow ring that surrounds the fabric bag. A relatively high velocity air stream is blown into the ring, and then out through a series of small holes located on the inside ring surface as the ring travels slowly up and down the outside of the bag. By this method, one section of the bag is cleaned at a time.

Many different types of fabrics are used in collectors, depending upon the particular conditions. The original fabrics were cotton and wool (prior to 1946, they were the only types available). Even today, cotton is used more than any other type (in terms of total yardage). The suitability of cotton and wool is very limited, and can be used only below 200°F.

After World War II, the polyamides (the nylons) came into use. They are very good in terms of abrasion resistance.

The acrylics entered the market in 1953. They are quite resistant to acid mist and can operate at higher temperatures than could the previously introduced fabrics.

In 1955, the polyesters were introduced. They are today the most widely used of the synthetics. They have good high-temperature stability and better abrasion resistance than the acrylics.

Fiberglass, introduced in 1956, is the only fabric that can be used above 500°F. However, it has poor abrasion resistance and cannot be mechanically shaken, since that would break the fibers.

Teflon (polytetrafluoroethylene, or PTFE) and nomex (poly-m-phenylene isophthalamide) were added in 1964. Teflon is quite good at high temperatures, (500°F) and is acid resistant, but it has the lowest coefficient of friction of all the materials; thus, the particulates may pass through the fabric when it is not desired that they do so. Teflon also has poor dimensional stability and low abrasion resistance, yet its cost is very high. Nomex is used primarily in the mining industry. It has good high-temperature (400°F) and reasonable abrasion resistance, but it is necessary to maintain low sulfur levels and its cost is also very high.[2]

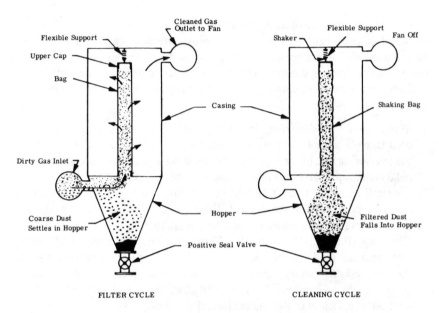

FIGURE 3-9. Shaker-type baghouse. (From *Air Pollution and Industry*, edited by R. D. Ross, © 1972 by Litton Educational Publishing, Inc. Reprinted by permission of Van Nostrand Reinhold Company.)

FIGURE 3-10. Shaker-type baghouse dust collector with two modules.

50

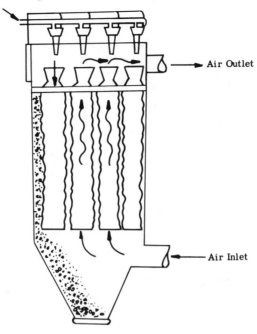

Air Jet for Cleaning

Air Outlet

Air Inlet

FIGURE 3-11. Air jet for baghouse cleaning. (Adapted from material appearing in the *1972 Report on Business and the Environment*, edited by Price, Ross, and Davidson, a McGraw-Hill publication.)

FIGURE 3-12. Pulse-type baghouse collector with two modules.

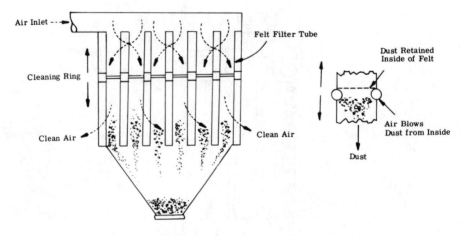

TRAVELING BLOW RING

FIGURE 3-13. Traveling blow ring used to clean baghouse. (Adapted from material appearing in the *1972 Report on Business and the Environment*, edited by Price, Ross, and Davidson, a McGraw-Hill publication.)

These various fabrics can appear in two major forms: as woven fabrics or as felted fabrics. The woven types can be as taffeta, twill, or sateen; other variations include yarn size, napped or unnapped, yarn ply, thread count, and yarn twist. The felted fabrics are more expensive. It is not possible to make a felted fabric from cotton or fiberglass.[3]

During the early to mid-1980s, baghouse use declined. By the late 1980s, however, the demand again rose, primarily due to additional fibers that could be used for new high-temperature ($\sim 900°F$) applications, such as recovering fly ash from fluidized-bed combustors and trapping dust in catalyst regenerators.

Nextel 312, an alumina/boria/silica fiber, can be woven, knitted, or braided to produce a fabric very similar in appearance to fiberglass. This material has been successfully tested at temperatures up to 2100°F in the laboratory. In the field, among other applications, it has exhibited 99.995% removal of particulates from a coal-fired boiler at temperatures of 850–900°F.

A fabric consisting of a membrane of expanded PTFE (teflon) laminated to a substrate of polyester or fiberglass is also growing in popularity. This fabric, known as Gore-Tex, is used on the feed side of the bag and is capable of efficient filtering immediately, without having to wait for a filter cake to build. The reported collection efficiency is > 99.9% of 0.3 μm diameter particles.

Other new fabrics include Chem-Pro, based on an aromatic polyimide fiber, P84, and Tyton, made from polyphenylene sulfide fibers.

The price of the different fabrics vary significantly and depend on factors such as baghouse size, surface finish, and site location. Industry estimates as to the relative costs, assuming woven fiber glass to be 1, are:[4]

Woven fiberglass	1
Polyester	0.75
Teflon	6–8
Gore-Tex on fiberglass	5–8
P84	3–4
Ryton	1.5–2
Nomex	1–2

In addition to the development of new fibers and fabrics, equipment manufacturers have, in recent years, refined all three baghouse designs to provide wider operating ranges, higher efficiencies, lower pressure drops, and longer life.

A related filter design that gained in popularity during the 1980s is the cartridge or pleated filter. Cartridge filters were originally used as a prefiltration device in gas-turbine intakes, but now their use has spread as dust collectors in the process industries.

Cartridge filters contain pleated filters, which were originally made of paper and hence considered rather fragile. Now the filters are available in other fabrics, such as Nomex or polyester. They are relatively small devices, inexpensive, and easily changed; however, they are limited to low dust loads and (often) low-temperature operation.

WET SCRUBBERS

Wet scrubbers use a stream of water to increase the collection efficiency. The solid particulates interact with this water stream and are removed as a slurry, or possibly a solution. If proper caution is not taken, this method could thus lead to a water pollution problem. In general, wet collectors cool and wash the air stream; some also remove gases. Because of the use of water, the permissible temperature and moisture content of the incoming dirty air stream is not limited. To further increase their versatility, it is possible to employ liquids other than water, which could possibly neutralize the corrosive nature of many air pollutants, such as SO_2. An additional advantage of wet scrubbers is the minimization of any explosion hazard due to dry dust and air mixtures. Several very costly explosions of this type have occurred in recent years, particularly in grain mills. Other advantages include a great variety in the types of scrubbers to suit a variety of situations, and a relatively low initial cost. The disadvantages include high operating cost, especially for the high-efficiency systems; possible mechanical failures due to dust buildup, erosion, and corrosion; required treatment of the scrubbing liquids and then disposal of the wet sludge; and the presence of a highly visible white plume, caused by steam, discharged to the atmosphere (Figure 3-14). The plume, though usually not harmful, can be a hazard to automobile and other types of traffic if a temperature inversion occurs by its creating an area of dense fog.

Corrosion is a major potential problem associated with wet scrubbers. To minimize the possible corrosive effects, they can be constructed of a variety of materials, such as stainless steel or other corrosion resistant materials, or they can be lined with plastic or rubber coatings.

There are two major types of wet scrubbers: 1) low-energy, and 2) high-energy (or venturi) designs. There are many varieties of each type that can be employed, and only a few typical samples will be considered.

Low-energy Types

Spray Chambers

One of the simplest types of wet scrubbers is a gravity settling chamber, with added sprays (Figure 3-15). The dirty gas enters the collector, and is slowed by the sudden volume expansion as it enters the main chamber. Simultaneously, the gas hits the water spray, which generates much turbulence. At this stage, many of the larger particles will settle due to gravity, inertia, and centrifugal force: the dust particles and water droplets collide and stick, and the resulting larger particles can settle fairly easily. The more the mixing, the more likely it is that the finer particles will settle. Higher-pressure sprays will produce more and finer water droplets, and more turbulence, and hence provide better collection efficiency. The remaining dust that does not settle in this area (in general, the smaller particles) then travels with the air stream into the de-mist section. Baffles in this section change the direction of air flow, further increasing the inertia and gravity effects, and forcing even more of the droplets to settle to the bottom. The addition of baffles also to the spray section can likewise increase the collection efficiency of the collector.

The water that is required for this type of process can be recycled after, for example, filtration, or the effluent can be discharged to a settling pond and fresh water can be used for the scrubber. This latter is practical only if the plant is located near a source of water, such as a lake or river.

Spray Towers

Spray towers are typically one of two types: open spray towers or packed towers. Open spray towers (Figure 3-16) are best for coarse particulates (>10 μm in diameter) such as those attained in the iron pyrite roasting step of steel production. They tend to be good for heavy particulate loads. A modification of this type of scrubber would require the water spray to be directed upward. The tower would be fitted with one or more baffles, each consisting of a perforated plate with hundreds of small holes/ft^2 and wetted with a thin layer of water. Partially across each of these holes is a

FIGURE 3-14. Water vapor generated by use of wet scrubbers for particulate collection. (*Photo courtesy DOE.*)

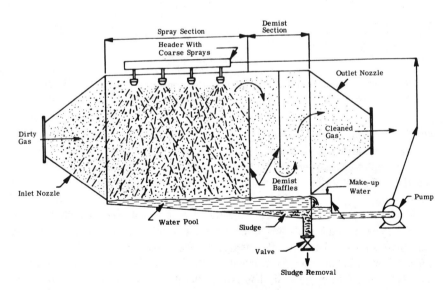

FIGURE 3-15. Gravity settling chamber scrubber. (From *Air Pollution and Industry*, edited by R. D. Ross, © 1972 by Litton Educational Publishing, Inc. Reprinted by permission of Van Nostrand Reinhold Company.)

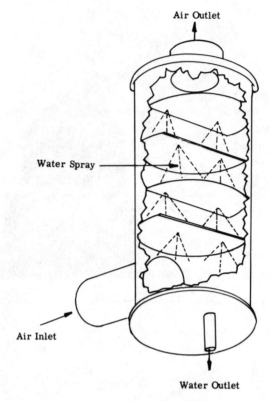

SPRAY TOWER

FIGURE 3-16.   Open spray tower. (Adapted from material appearing in the *1972 Report on Business and the Environment,* edited by Price, Ross, and Davidson, a McGraw-Hill publication.)

small impingement plate (Figure 3-17). The dust and water spray travel upward, through the baffle holes, where the liquid layer is partially atomized into about 10-μm droplets. These very small droplets are particularly efficient in collecting the fine dust. The water-dust droplets strike the impingement plate, settle, and drain out the bottom.

A packed tower (Figure 3-18) consists of a series of contact beds, through which the gases and liquids flow either countercurrent or crossflow. Since the gas flow path is long and the area of wetted surface is large, there is much opportunity for the dust to be captured by either the wetted surface or the water flow itself. This type of collector typically has little solids-handling capacity, and hence is not suitable for heavy dust loads. It is, however, suitable for gas cooling and absorption. A typical application would be for the absorption of fluorine compounds.

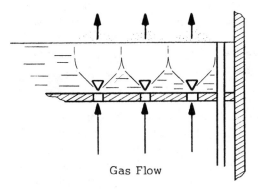

Water Droplets Atomized
At edges of Orifices

Gas Flow

Downspout to Lower Stage

FIGURE 3-17. Impingement plate on open spray tower.

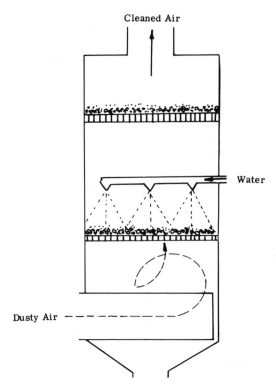

Cleaned Air

Water

Dusty Air

FIGURE 3-18. Packed spray tower. (From *Pollution Control Technology*. New York: Research & Education Association.)

Higher dust loads can be handled if the packing material consists of floating spheres. In a "fluidized bed scrubber" (Figure 3-19), the gas flow rate is sufficient to keep the spheres suspended and in constant motion. This motion prevents plugging of the bed and the possible development of particular air flow channels that would decrease the collection efficiency.

Other Types

Other possibilities of low-energy scrubbers include cyclonic types (based on centrifugal force), flooded bed scrubbers (similar to fluidized bed in operation, but with lower gas flow rates and thus less motion and no suspension of the packing material), and submerged-orifice (or nozzle) scrubbers, where the dirty gas makes contact with a sheet of water in a restricted passage. Typically, these types of collectors are fairly efficient (85–95%) on a minimum particle size of 2–5μm.

High-energy Types

High-energy wet scrubbers (venturi scrubbers) are more efficient than are the low-energy varieties. In the venturi scrubbers, the dirty gas and scrubbing liquid are moving at very high velocities (100–450 mph), generating much turbulence and mixing, and thus exhibiting good collection efficiency.

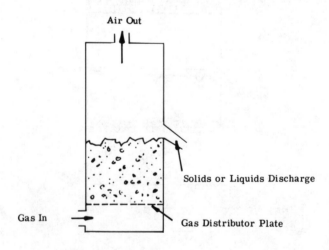

FIGURE 3-19.   Fluidized bed scrubber.

There are two basic types of venturi scrubbers: dry and wet. The dry types (Figure 3-20) are not really dry; the first section, where the dirty gas enters, concentrates, and speeds up due to the narrowing passageway, is dry, but as the gas passes through the constricted area, it is inundated by a transverse water spray. The wet types, on the other hand, require that both collector sections be wet. The water spray contacts the dust as soon as it enters the collector and the two mix immediately. One advantage of this immediate mixing is to allow collection of hotter gases.

With venturi scrubbers, there are many variations as to how the water and dirty gas are injected, and how the sludge and mist are removed. Flooded disc scrubbers (Figure 3-21), for example, inject the water via a disc that can be adjusted to vary the air flow rate for maximum collection efficiency. A jet scrubber frees the water under such high pressures as to aspirate the dirty air, and thus works well to remove mist or easily absorbed gases. Multiple venturis, set in parallel, are also possible.

Venturi scrubbers are very efficient (better than 99%) on particles as small as 0.5 μm. As such, they are typically used in the steel industry for fine particulates,

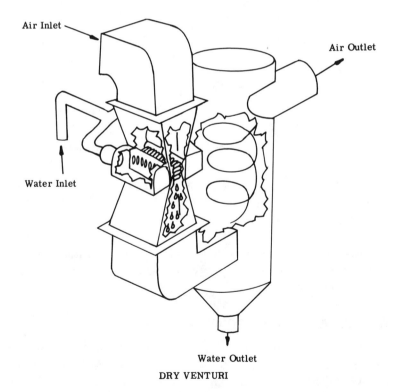

Air Inlet

Air Outlet

Water Inlet

Water Outlet

DRY VENTURI

FIGURE 3-20.   Dry venturi scrubber. (Adapted from material appearing in the *1972 Report on Business and the Environment*, edited by Price, Ross, and Davidson, a McGraw-Hill publication.)

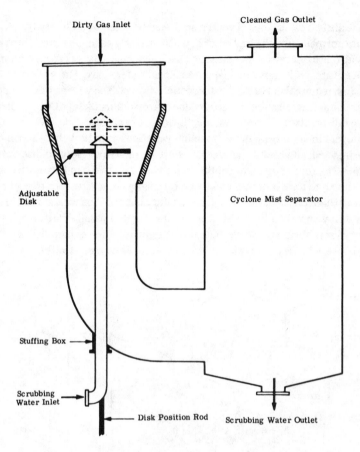

Dirty Gas Inlet

Cleaned Gas Outlet

Adjustable
Disk

Cyclone Mist Separator

Stuffing Box

Scrubbing
Water Inlet

Disk Position Rod   Scrubbing Water Outlet

FIGURE 3-21.   Flooded disc scrubber. (From *Air Pollution and Industry,* edited by R. D. Ross, © 1972 by Litton Educational Publishing, Inc. Reprinted by permission of Van Nostrand Reinhold Company.)

for acid mists, and for lime and coal dust (whose size is often 1 μm or less). The larger water pressure drops of the high-energy systems do lead to relatively high operating cost, however.

ELECTROSTATIC PRECIPITATORS

The electrostatic (or Cottrell) precipitator is a very versatile and efficient type of collector, which can operate at over 1000°F, and can be used for dry particulates or fumes, as well as mists. Electrostatic precipitators operate on the basis of electrostatic attraction. The dirty gas is channelled between two electrodes, a high-voltage (between −40,000 V and −50,000 V) discharge electrode and a grounded collecting electrode (Figure 3-22). The high negative voltage causes electrons to be emitted from the discharge electrode; these electrons then collide with the surrounding gas molecules, ionizing them. The negative gas molecules migrate to the relatively positive (grounded) electrode and, in doing so, collide with and transfer their negative charge to the entrained dust or mist particles. The negative dust or mist particles then also migrate to the grounded electrode, where they become neutralized. Mists will simply run off, but the dust particles often adhere to the plate. As more of a dust layer builds on the collecting electrode, the neutralization efficiency of that electrode decreases due to the presence of the insulating dust. It is necessary to frequently rap the plate with hammers, or to vibrate it, to prevent any major buildup of dust and to keep the precipitator operating. The dry particles then drop to the bottom hopper for removal.

Usually, a precipitator will have more than one set of electrodes, and thus the dust can be collected in different "compartments."

Precipitators allow for the dry collection of very fine particles. The efficiency,η, will, of course, vary with the particle diameter and various operating parameters. It is commonly estimated by an empirical equation:

$$\eta = 1 - \exp\left(\frac{-AV_d}{Q}\right)$$

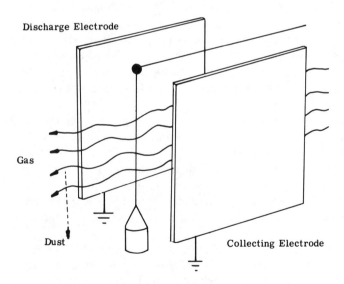

PRECIPITATOR

FIGURE 3-22.   Electrostatic precipitator. (Adapted from material appearing in the *1972 Report on Business and the Environment,* edited by Price, Ross, and Davidson, a McGraw-Hill publication.)

where, A = area of collection electrodes, m^2
   Q = volumetric flow rate of gas, m^3/s
   V_d = drift velocity, m/s

The drift velocity is the velocity of the particulates to the collecting electrode. Generally, it is assumed to be 0.5 times the particle diameter, when expressed in μm. Typical drift velocities are 0.03–0.2 m/s.

Precipitators are highly efficient, often approaching 99.9% efficiency. They are also good for hot and/or corrosive substances, although, because of high installation costs, it may be preferable to use a different collection technique, if possible. Operating and maintenance costs are low compared to other high-efficiency collectors, however.

The conductivity of the dust is also a limiting factor in the use of precipitators. The resistance of the dust layer that does develop, the "dust resistivity," is preferentially in the range of 10^7–10^{11} Ω/cm. If the dust has too low of a resistance, it may reenter the air stream by too readily transferring its charge to the plate; if the resistance is too high, it would have too great a tendency to adhere to and insulate the collection electrode. This would reduce the effective potential across the electrodes and possibly lead to spark discharges, reversing the ionization, and thus also causing reentry of the dust to the air stream. The resistivity of dust such as that generated by low-sulfur coal can be reduced by the intentional addition of particular substances to the air stream: moisture, ammonia, acid mists, and SO_3 are possibilities.

Spark discharges, with the subsequent decrease in collection efficiency, also can be caused by dust loads that are too high. The equipment should not be operated above its design capacity. Another problem, experienced by several companies, is the possibility of the precipitator getting too hot. This has led in several instances to the precipitator igniting and the steel actually burning out of control, and totally destroying the precipitator.

Normally, maintenance is inexpensive. The typical problems encountered are discharge electrode failures and rapper malfunctions. As with all other high-voltage electrical equipment, there are possibilities of transformer rectifier failures. Though the dust collected in the hopper is theoretically electrically neutral, in reality it usually has some residual charge. One cement plant, for example, noted that the type of charge varied with weather conditions. This residual charge and, thus, the mutual repulsion of the particulates, encourages hopper plugging.

There are two types of precipitators: plate varieties and pipe varieties. The plate types are the more common and usually are used on dry dust. They consist of a series of flat collecting plates, with the high-voltage discharge wires suspended between them. The grounded collecting plates range from 3–9 ft long and 10–36 ft high. They are grouped together within a completely enclosed casing, and are independently suspended and rapped for cleaning. The discharge wires are electrically insulated from the remainder of the system, and usually consist of 12-gauge steel spring wire, reinforced at the two ends for good electrical contact and resistance to wear.

Pipe varieties are better for liquid aerosols, sludge, or fumes. Usually, a number of grounded pipes, grouped together in a cylindrical casing under a header plate, are used as the collector electrodes, and the discharge wires are suspended within each pipe. The liquid wastes can then be guided by the pipes as they run out of the system. There is a wide variation in capital costs, and even operating costs are dependent upon the proposed applications.

FUME INCINERATION[5]

Fume incineration is used primarily for odor removal. It also has some possibilities for the control of some hydrocarbons and photochemically active substances.

There are three methods of fume incineration:

1. Direct flame incineration
2. Thermal incineration
3. Catalytic incineration.

FIGURE 3-23.   Many industries, particularly those related to petrochemicals, still flare their low-concentration organic emissions, burning them open to the atmosphere with an almost invisible flame.

Direct flame incineration is usually used when the waste materials will readily combust when mixed with air. If the dirty air stream itself does not contain enough combustibles to sustain combustion, air and a small amount of natural gas may be added to increase the combustibility. This mixture can then be burned in a simply designed flare-type incinerator. An example of a gas suitable for incineration is hydrogen cyanide (HCN). HCN is an extremely poisonous, colorless gas evolved when cyanides are treated with acids and used in the production of, for example, nylon. Incineration will destroy the toxicity, producing a product (CO_2, N_2, and H_2O) that can be released to the atmosphere.[6] The typical temperature reached is about 2500°. In petroleum refineries and petrochemical plants, the incinerator, or flare, is usually aimed upward and operates in the open, producing a bright light and significant noise. In other situations, the burner can be aimed horizontally and/or can be fired into some enclosure. In general, direct flame incineration is the simplest method of incineration.

Usually, the dirty gas stream does not contain adequate combustibles to support a flame without the addition of a large amount of fuel. This becomes unfeasible. Instead, a gas burner can be used to heat the air stream, typically to 1000–1500°F. At these temperatures, the pollutant fumes will undergo thermal degradation. Usually, the gas stream will have a residence time in the burner (Figure 3-24) of about 0.5 second.

Catalytic incineration typically employs platinum or paladium catalysts for the decomposition of paint solvents, and odors arising from lithographing, food preparation, and similar processes. Often, an alumina support, in the shape of a wire screen, air foil shaped rods, spherical or cylindrical pellets, or a honeycomb, has

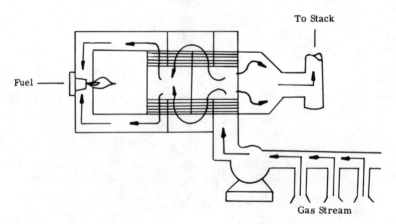

FIGURE 3-24.   Thermal incinerator. (Adapted from material appearing in the *1972 Report on Business and the Environment,* edited by Price, Ross, and Davidson, a McGraw-Hill publication.)

dispersed on its surface the catalyzing metals. The shape of the support allows for a large contact area between the gas stream and the catalyst. Generally, the catalytic reactions are carried out at 600–1000°F. When using catalytic incineration, it is necessary to take precautions so that the waste gases have no catalyst "poisons" in them; materials such as phosphorus, silicon, lead, iron, vanadium, and arsenic tend to poison many catalysts, destroying their ability to function.

SORPTION TECHNIQUES

Volatile materials and odors can also be removed from the effluent by absorption into liquids and absorption onto the surface of solids.

Adsorption frequently employs materials such as activated carbon, silica gel, or alumina, all solid substances with a very large surface-to-volume ratio. Generally, the process is used for removing volatile organic carbon (VOC) compounds, though it is also suitable for sulfur-bearing materials and mercury. The adsorptive capacity of the solid depends on both the properties of the solid and those of the organic, as well as factors such as temperature and pressure. Generally, the adsorptive capacity increases with molecular weight of the material adsorbed, with its degree of cyclization, and with its polarity.

Absorption, on the other hand, typically utilizes either water or an organic liquid such as a low-volatility hydrocarbon or mineral oil. Absorption occurs whenever the concentration of the organic species in the liquid is less than the equilibrium concentration. Thus, the amount absorbed is a function of both the physical properties of the system and the specific operating parameters.

Absorption techniques can be used in conjunction with spray chambers, packed towers, and venturi scrubbers, to enhance the removal efficiency of volatile substances. Maximum absorption occurs at low temperatures, by using large contacting surface areas, high liquid-to-gas ratios, and concentrated gaseous streams. For inlet concentrations of 5000 mg/L, removal efficiencies >98% may be achieved.[7]

TALL STACKS

The purpose of tall stacks is to diffuse the pollutants, often SO_2, not to reduce the quantity. As a consequence, the results obtained by using this technique are mixed. For example, dispersing SO_2 will minimize its concentration in the immediate vicinity of the source, which is very beneficial due to its innate toxic effects. However, dispersing the SO_2 over too great an area will also increase its residence time in the atmosphere, potentially leading to an increase in acid deposition at many miles distant.

The SO_2 and NO_x emitted directly from the stack are called the "precursor gases." In the presence of sun and other atmospheric catalysts, the precursors are

FIGURE 3-25.   Tall stacks remain one method of choice for dispersing atmospheric pollutants over large areas. Sulfur oxide emissions, especially, are frequently handled at least partially by this technique. This is the 1250-ft tall stack belonging to International Nickel Company of Canada, Ltd.

slowly converted to "acid formers," which are solid particles containing SO_4^{2-} and NO_3^- ions. When these acid formers then combine with moisture, they form H_2SO_4 and HNO_3. Acid deposition is a combination of "dry-fall," the precursor gases and acid formers, and "wet-fall," the acids themselves.

Though there does not appear to be a direct correlation between either emission rate or stack height and acid rain impacts downwind, it is good engineering practice to design a stack for only moderate dispersal of the pollutants. As a rule of thumb, the stacks should be 2½ times the height of the tallest nearby building.

Recent studies, however, have indicated that the stacks might not be as effective in dispersing the pollutants as once thought. Gerald De Marrais, of the National Oceanic and Atmospheric Administration, reported that the air masses at 100 m are much more stable than had been assumed. This means the effluents get less mixing and dispersion than projected.[8]

Even without this new information, the stacks often had to be so tall that some other method of control would turn out to be cheaper. For example, International Nickel Company of Canada, Ltd., has a 1250-ft stack at Cooper Cliff, Ontario, which cost $5.5 million when installed in 1972. This $5.5 million might have been better spent if it had been invested in converting the hot smelter operations to the new chemical leaching techniques available for processing.[9]

References

1. "...Don't Control Pollution, Prevent It." *Environmental Science and Technology* **11**, March 1977, p. 234.
2. Schoeck, V. E., "In Mechanical Dust Collectors, It's The Fabric That Counts." *Business and Environment*, Fred C. Price, Steven Ross, and Robert L. Davidson (Eds.). New York: McGraw-Hill, 1972, pp. 4-16.
3. Ross, R. D., *Air Pollution and Industry*. New York: Van Nostrand Reinhold, 1972, p. 364.
4. Parkinson, Gerald, "A Hot and Dirty Future for Baghouses," *Chemical Engineering* **96** No. 4, April 1989, pp. 30-35.
5. Cheremisinoff, Paul N., "Techniques for Industrial Odor Control." *Pollution Engineering*, October 1975, p. 24.
6. Ross. p. 453.
7. McInnes, Robert, Jelinek, Steven, and Putsche, Victoria, "Cutting Toxic Organics." *Chemical Engineering* **97**, No.9, September 1990, pp. 108-113.
8. "Very Tall Stacks Not Too Useful." *Chemical and Engineering News*, February 20, 1978, p. 23.
9. "Air Pollution Control." *Business and Environment*. pp. 4-8.

4

Methods for Water Pollution Control

Manufacturing is the leading source of controllable man-made water pollutants; domestic waste is second. As mentioned previously, the industrial wastes are more likely to contain substances that will resist the normal treatment procedures. However, except for the types of industries that generate large amounts of incompatible wastes, many industries take advantage of the local municipal treatment facilities for some or all of their waste waters. It is often necessary to pretreat some of the industrial wastes before introducing them into the municipal treatment facility, but most of the time, there is no difficulty in handling the industrial wastes by the normal municipal technologies. Industrial wastes frequently comprise a large percentage of the volume treated by municipalities. For example, the treatment facility in Green Bay, Wisconsin was originally financed jointly by the Green Bay Metropolitan Sewer District, the American Can Company, and the Procter and Gamble Paper Products Company. Other local industries also use the facility, paying a sewer bill as do the residential users. A total of $> 60\%$ of the BOD is industrial in origin at this particular facility. Even if the industry has its own treatment facility, it usually operates on the same principles and employs many of the same techniques as the municipal systems.

There are three broad classes of treatment methods that are employed:

1. *Primary treatment.* Primary treatment removes from the wastewater those substances that float or will settle out. All processes in this category concentrate on removing the pollutants by physical means; hence, this constitutes the mechanical treatment stage. Techniques included are grit removal, screening, grinding, and sedimentation.
2. *Secondary treatment.* This step is always based on biological oxidation; thus, it is a reproduction of the degradation processes that occur in nature. The major

68

FIGURE 4-1.   Open flow of wastewater to sewage treatment facility, Vienna, Austria.

purpose is to remove the soluble BOD, as well as the suspended solids that were not removed in the primary treatment. The three common methods used are activated sludge, trickling filters, and oxidation ponds (lagoons). All are based on having various microorganisms feeding on the organic impurities in the presence of O_2, at a favorable temperature, and for a sufficient time period.

3. *Tertiary treatment.* This is the most advanced treatment. Included in this stage are primarily the various chemical treatments of waste water.

Most treatment facilities include primary and secondary treatment; some, particularly those associated directly with an industry, also include some form of advanced treatment.

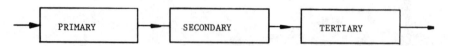

FIGURE 4-2.   Schematic of typical overall wastewater treatment process.

There are many variations as to what processes are employed, but basically we find the waste water is treated as in Figure 4-2.

PRIMARY TREATMENT[1]

A typical primary (or mechanical) treatment stage is shown in detail in Figure 4-3. Most of the objects removed during the initial bar screening step (Figure 4-4) are quite large, and vary from logs to tires to anything else that makes its way into the sewer systems. There have even been instances when the corpse of a dead baby was found. These screens are typically parallel steel or iron bars with approximately ½-in openings. Often, these screens are placed in an inclined position, or can rotate on a horizontal axis, for ease of cleaning. The debris is usually landfilled.

After screening, the water goes to a grit remover (Figure 4-5), where sand, small stones, and cinders are allowed to settle. After washing, this grit is also landfilled.

Next, a primary settling basin (Figures 4-6 and 4-7), equipped with a skimmer, removes most of the oil and grease as well as much of the heavier settleable organics. These tanks are usually 10–12 ft deep and retain the water for 2–3 hours. The solids that settle, called the raw primary sludge, are removed by mechanical scrapers and pumps. The water containing the nonsettleable pollutants then flows by gravity to the biological treatment process. Most of this pretreatment is not required for industrial wastes, or the pretreatment is done at the plant, so those wastes can be shunted directly to the biological steps.

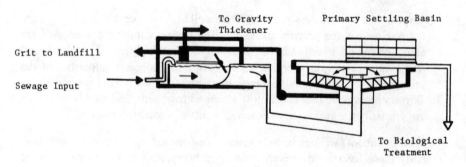

FIGURE 4-3.   Schematic of typical primary treatment process.

FIGURE 4-4. Preliminary for screening of waste water to remove large objects. (*Photo courtesy of John Kirchner and the Green Bay Metropolitan Sewerage District.*)

FIGURE 4-5. Grit remover. Note conveyor system, which allows settling of sand and dirt particles. (*Photo courtesy of John Kirchner and the Green Bay Metropolitan Sewerage District.*)

FIGURE 4-6. Primary settling basin. The oils and greases are skimmed from the top; the solids settle. (*Photo by David Garot.*)

FIGURE 4-7. Overview of the primary settling basins. (*Photo courtesy John Kirchner and the Green Bay Metropolitan Sewerage District.*)

SECONDARY TREATMENT

There are three common procedures for secondary treatment: activated sludge, trickling filters, and oxidation ponds.

Figure 4-8 illustrates a secondary process employing an activated sludge treatment. The effluent enters aeration basins (Figures 4-9 and 4-10). Microorganisms and air are mixed with the effluent by means of agitators; the bacteria, protozoa, algae, fungi, or whatever grow and multiply, using the impurities as food, and simultaneously breaking down the organics present. As the microorganisms grow, they tend to clump together, forming the "activated sludge." After several hours, the mixture of waste water and activated sludge flows to settling basins (Figure 4-11). Some of the bacteria that settle in the sludge are taken to reaeration basins, where they are kept alive and given air but no food; they become "lean and hungry," suitable for use again in the aeration basins. The remainder of the sludge is taken to air flotation areas for further treatment. The effluent is taken to chlorine contact basins, where chlorine is added to kill any disease-producing organisms, and then is discharged to the nearby natural water systems or to further treatment.

Another common biological treatment is a trickling filter. This method consists primarily of a 10-20-ft deep, 200-ft diameter coarse filter bed. The effluent is spread out over the bed in drops, films, or spray from either moving distributors or fixed nozzles. The waste water trickles through the bed to under-drains. Retention time in a trickling filter can be as long as 18 hours, though generally it is less.

Microorganisms grow on the gravel surfaces, and act on the dissolved organics as they filter downward with the waste water. The rocks are typically 1-4 inches in diameter, so they provide fairly large amounts of surface area on which the microorganisms can grow. As the microorganisms grow and reproduce, some do wash out of the rock media, so it is necessary to have the flow from the filter passed

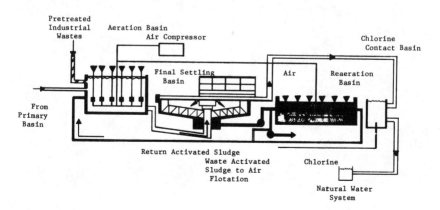

FIGURE 4-8.   Activated sludge process.

FIGURE 4-9. Activated sludge process. (*Photo courtesy of John Kirchner and the Green Bay Metropolitan Sewerage District.*)

FIGURE 4-10. A view of the empty activated sludge basin, showing the aerators. (*Photo courtesy John Kirchner and the Green Bay Metropolitan Sewerage District.*)

FIGURE 4-11.   Activated sludge settling basin. (*Photo by David Garot.*)

through a sedimentation basin to allow the organisms to settle out. High organic loads do tend to plug trickling filters. To try to minimize this, rocks have in some cases been replaced by corrugated plastic sheets, plastic rings, and redwood slats in an effort to provide even more surface area. The efficiency of this type of system is not quite as good for removing BOD as is the activated sludge procedure, typically removing only about 85% of the BOD. The procedure is affected greatly by air temperature (being less efficient in winter), but the power consumption is low, the mechanical equipment is simple, and the incoming pollutant load can vary over a wide range without causing any problems. Activated sludge, on the other hand, can remove essentially all of the BOD and is lower in capital costs than a trickling filter. However, the process requires more energy—the air compressors are the most energy-intensive part of the plant—and also more operational control.

The third type of treatment facility is an oxidation pond, or lagoon. Oxidation ponds are large, shallow ponds designed to treat waste water through the interaction of sunlight, wind, bacteria, O_2, and algae. Ponds are usually 2-4 ft deep, shallow enough to allow mixing by wind currents and to have O_2 present in all except the bottom (sludge) layer. Some ponds are deeper, 10-20 ft deep, and have a much greater anaerobic (O_2-depleted) zone. The algae grow using sunlight and the CO_2 and inorganic substances released by bacteria in the pond, simultaneously releasing O_2 for use by bacteria. The bacteria, as in the other methods, feed on the dissolved organics.

Aeration equipment can also be used to supply O_2 to the pond, forming a type of activated sludge process.

Though the ponds are easy to construct, maintain, and operate, they do require large amounts of space, limiting their applicability to smaller municipalities or to individual plants, such as some paper plants.

SOLIDS PROCESSING

Figure 4-12 shows just one possible solids processing procedure. The waste activated sludge first enters an air flotation section. Air is blown into the sludge. As the bubbles rise, the solids are also borne up to the top of the tank, where they are skimmed off and thus thickened. The sludge from the primary settling basins (part of the mechanical treatment) is taken to a gravity thickener, where it is further thickened. The solids that settle in this tank are taken to a sludge holding tank, as are those from the air flotation. From there, the solids go to the thermal conditioning unit. The thermal conditioning unit acts as a pressure cooker and digests the sludge. The high heat and pressure forces the sludge to release the majority of the trapped water. Then the sludge is taken to a vacuum filter (Figures 4-13 and 4-14) and dried to about 35% solids. The vacuum filter can

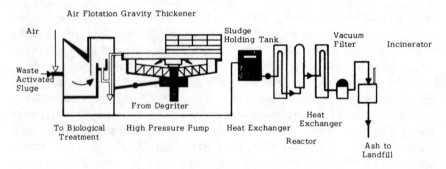

FIGURE 4-12.   One possible solids processing procedure.

FIGURE 4-13.   Vacuum filtration. (*Photo by John Kirchner and courtesy of the Green Bay Metropolitan Sewerage District.*)

consist of a large, slowly rotating roll, the surface of which is spread with sludge. As the rolls turn, a vacuum is applied through the roll surface and the sludge dries and forms a cake, which then falls off. The semi-solid sludge is then sent to an incinerator. The organic content is typically great enough to sustain combustion without the addition of supplementary fuel. The resulting ash is finally disposed of in a landfill.

There are many other possible methods of treatment. Often, somewhere in the procedure (for example, in the primary settling tanks), flocculation can be instituted to help the fine solids settle. One possible flocculation technique is to add potassium aluminum sulfate, $KAl(SO_4)_2$, and then NH_3. The two will react, forming aluminum hydroxide, $Al(OH)_3$, a gelatinous precipitate that remains suspended. This gelatinous suspension will then physically gather other suspended materials and incorporate them into its structure. The $Al(OH)_3$ and other suspended materials can then be filtered out together, leaving relatively clean water. This technique works only for suspended solids, however, not for the dissolved species. Other chemicals that can be used include ferric chloride and lime.

Many of the suspended solids are very small, even colloidal in size, and hence are very difficult to remove. They typically possess negative charges. Precipitation thus requires 1) charge neutralization (coagulation), and then 2) the formation of clumps, or flocs (flocculation).

FIGURE 4-14.   The screen of the vacuum filtration unit. (*Photo by David Garot.*)

The normal explanation for coagulation is the "double-layer model." The small particles, negatively charged in solution, attract oppositely charged counterions. The attractive forces are sufficiently large, so the inner layer of positive ions migrates through the waste water attached to the particles, forming what is called a slippage plane. Additional counterions are loosely attached in an outer layer, but these can easily be removed. The result is a negatively charged ion, whose negative charge is partially reduced by the positive ions in the inner layer. The net charge of this complex is called the zeta potential. This is, essentially, the force that prevents neighboring particles from approaching each other.

The purpose of coagulation is to decrease this mutual repulsive force to zero. The most common approach is to add trivalent cations. These trivalent ions, because of their greater charge, can replace any monovalent ions in the inner layer. As a result, they can lower the net effective charge of the complex, and decrease the repulsive forces.

Flocculation is encouraged by the development of velocity gradients in the wastewater, often by a rotating paddle. If various portions of the waste stream travel at different speeds, collisions are enhanced, enhancing floc growth.

The sludge handling and disposal steps are very important and also very expensive, representing 25-50% or more of the total treatment costs. It is really the single most troublesome step of the treatment procedures today.

The sludge is a semi-liquid whose initial solid content varies, depending upon the source. That from the primary settling basins is typically 2.5-5% solids; from trickling filters, 0.5-5% solids; and, from the activated sludge procedure, 0.5-1% solids. The sludge often contains water internally in its physical structure, and this water is very difficult to remove.

The objectives of all types of sludge treatment are several:

1. To convert the organics to some stable form;
2. To reduce the volume by removing liquids;
3. To destroy harmful organisms; and
4. To obtain byproducts, if possible, to defray the overall costs.

To accomplish these objectives, several general methods are employed.

Concentration

Concentration of the sludge can be accomplished by 1) "clarifier thickening," or by 2) separating the liquids from the more solid portions. Clarifier thickening basically means flocculation; it can be accomplished by the addition of special chemicals, as discussed earlier, or it can occur naturally. Many microorganisms have the ability to naturally flocculate and later separate in the system. The procedure often involves organic and biochemical polymers, as well as inorganic substances.

Separation of the more solid portions can be done by several techniques. Gravity thickening (Figure 4-15) is one possibility. The efficiency of this type of system is inversely proportional to the rate of overflow of the liquid and the basin depth; therefore, the basin should be as shallow as feasible and the liquid should flow slowly. Gravity thickening tanks are very similar in appearance to those used in the primary and secondary sedimentation. The solids that settle are scraped into a hopper, where they are collected for further processing. The best results are obtained with this system when only primary sludges are treated—no activated sludges. Primary sludges can generally be thickened from less than 5% solids to 10% solids by gravity thickening.

Flotation (Figure 4-16) is another possibility; the solids are floated to the surface of the liquid by small bubbles of air or biologically generated gases. For example, air can be injected into the sludge under pressures of about 40-80 psi. A large amount of the air can be dissolved at these pressures, air that later comes out of solution by forming minute bubbles that float to the surface. The sludge that attaches itself and floats up to the surface with the bubbles is then removed by a skimming mechanism. This method is particularly efficient on activated sludge and typically increases its solids content from 0.5-1% to 3-6%.

FIGURE 4-15. Gravity thickening basins for sludge concentration. (*Photo courtesy John Kirchner and the Green Bay Metropolitan Sewerage District.*)

FIGURE 4-16. Air flotation unit. (*Photo courtesy John Kirchner and the Green Bay Metropolitan Sewerage District.*)

Stabilization

The sludge is usually stabilized by some form of digestion to prevent any reaction of the more active ingredients after disposal. Digestion of the sludge can occur by using either aerobic or anaerobic bacteria. Aerobic bacteria require free molecular O_2. Ideally, the organic matter can be decomposed by these organisms to CO_2, H_2O, and relatively innocuous nitrates, phosphates, and sulfates. Anaerobic bacteria use the O_2 present in chemical compounds, and the decomposition tends to generate some disagreeable odors due to the formation of methane (CH_4), which can be used as fuel, ammonia (NH_3), or hydrogen sulfide (H_2S). The anaerobic process is also slower than the aerobic one.

Anaerobic digestion usually is carried out in two stages, in two deep (20–45 ft), large-diameter (115-ft) tanks. The biological activity is carried out in the first tank, the contents of which are heated and mixed. The second tank is used for storage of the digested sludge and the supernatant liquid that forms in the process. This liquid has a high concentration of soluble pollutants, and must be recycled to the treatment plant. The CH_4 generated can be used as a fuel to operate sewage pumps and blowers and to generate electricity.

Aerobic digestion is carried out in tanks similar to those used for the activated sludge process. It is as efficient as the anaerobic process, achieving about 50% conversion of solids to liquids or gases, and has the advantage of being more stable in operation and of recycling fewer pollutants back to the treatment plant. However, because CH_4 is not produced aerobically, this potential energy source is not an available by product. Aerobic digestion also requires more power.

Dewatering

Dewatering the sludge is commonly done by one of the following:

1. Drying beds
2. Vacuum filtration
3. Pressure filtration
4. Centifugation.

Drying beds have been a very popular means of sludge dewatering. The sludge is spread over a bed consisting of a 4–9-in layer of sand placed over 8–18 inches of gravel, and is allowed to stand until dried by a combination of evaporation and drainage. The water that does drain off is collected in pipes beneath the gravel and recycled to the treatment plant. By this procedure, the sludge can reach 45% solids in six weeks, and may eventually reach 85–90%, weather permitting. The dried sludge is then removed (by hand or mechanically). These drying beds are simple to operate, but they are practical only for small-volume systems due to the large area required. Unless the beds are covered, their performance also varies significantly with the weather conditions.

Vacuum filtration (Figure 4-17) consists of a cylindrical drum that is covered with some filtering material. The drum rotates partially submerged in a vat of sludge. As the drum rotates, a vacuum is applied from inside the drum. The vacuum extracts the water, leaving the sludge on the surface. Then the filter cake is either scraped off automatically with a blade, or it is dislodged by passing the fabric over small rollers. The sludge can, by this method, reach 30-40% solids, the process being less efficient on activated sludges. This method is very popular, particularly municipally.

Pressure filtration can also be effectively used for sludge dewatering. The sludge is pumped at high pressures through a filter medium. The solids are retained on the surface, and very clear liquid is filtered through. This method produces filter cake greater than 50% solids, though often the operation costs are high.

Centrifuges use centrifugal force to separate the liquids and solids. The sludge, often plus flocculating polymers, is pumped into a large, horizontal, cylindrical rotating container. As the container rotates at 1600-2000 rpm, the solids are forced to the outside, from where they are removed by a screw conveyor. The liquid is returned to the treatment plant; the sludge, 15-20% or more solids, must be disposed of. Usually, the solid content is not great enough for them to be incinerated without auxiliary fuel. Centrifuges are difficult to maintain because of the high speeds, but often they can handle sludges that vacuum filtration cannot, and at about the same operating costs. Another possible difficulty is an excessive buildup of fines in the liquid, and hence on the treatment facility to which it is recycled.

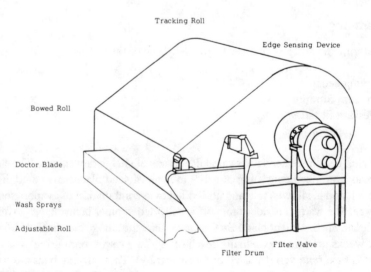

FIGURE 4-17.   Vacuum filtration.

Other methods of dewatering have also been tried. Milwaukee, Wisconsin, for example, has attempted to break down the sludge cell walls by freezing and thawing. Gamma irradiation has been done by Chicago, Illinois and appears effective, particularly in reducing the volume of waste-activated sludge. Simultaneously, the radiation also disinfects any possible harmful substance that might be present. Conditioning with high-energy electrons is another possibility, used at Deer Island, near Boston, Massachusetts.[2] Another method, of course, is the thermal conditioning "pressure cookers" previously discussed. Heat exchangers could be used in this situation to reduce the energy costs.

Heat Drying and Combustion

The sludge can then undergo one of the following three treatments:

1. Heat drying
2. Incineration
3. Wet oxidation.

Heat drying is usually done after mechanical dewatering, if at all. Not many facilities do it any longer, because there is usually not enough return from the fertilizer product to make it economically viable.

FIGURE 4-18.   City of Chicago—Northwest Municipal incinerator. (*Photo courtesy Wisconsin DNR.*)

Incineration really has two aims: to reduce the volume and to sterilize the organics. The two most common incineration methods are 1) the multiple hearth furnace, and 2) the fluidized-bed incinerator.

The multiple hearth furnace (Figure 4-19) typically consists of a circular steel shell that surrounds a number of hearths and is lined with refractory brick for insulation and protection of the shell. The dewatered sludge enters from the top, and travels downward from one hearth to another, being moved by a series of arms extending outward from the central shaft. Oil or gas burners provide supplemental fuel for startup and also, if necessary, for during operation. Figure 4-20 is a closeup of the interior of a multiple hearth furnace that is burning sludge.

The fluidized-bed incinerator consists of a vertical steel cylinder with a bed of sand. The combustion air enters from the bottom with enough velocity to fluidize the sand—to keep the particles in suspension in the air stream. The dewatered sludge is injected into the fluidized sand, where it is oxidized. The ash is carried out the top with the air stream, where it is collected.

Both incineration systems are comparable in cost. The multiple hearth furnace is easier to operate and maintain, but it requires much longer startup times to prevent

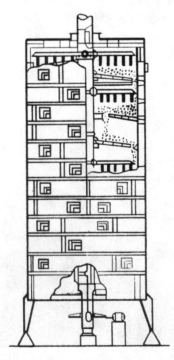

FIGURE 4-19.   Schematic of a multiple hearth furnace that can be used for incineration of sludge.

FIGURE 4-20.   The bottom interior of a multiple hearth furnace, being used to burn waste treatment sludge. (*Photo courtesy of John Kirchner and the Green Bay Metropolitan Sewerage District.*)

damage to the refractory bricks. The fluidized bed maintains heat in the sand, so startup times are shorter and the fuel is also used more efficiently. However, maintenance and operation are more difficult.

Wet air oxidation could also be used, particularly if the solid content of the sludge is quite low. The procedure is based on the fact that any substance that can burn can also be oxidized in the presence of water at 250–700°F. The sludge is first

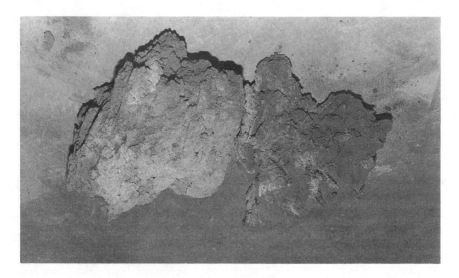

FIGURE 4-21.   Noncombustible incinerated metals. (*Photo courtesy John Kirchner and the Green Bay Metropolitan Sewerage District.*)

ground, and then put in a high-temperature and high-pressure reaction chamber (500°F, 1000–1700 psi). At these pressures, the water remains as a liquid. Air is injected to aid in the oxidation. The solids react with the O_2, and the solid products can then be separated from the liquid by settling in lagoons, vacuum filtration, or centrifuging. The liquid must be recycled to the treatment plant because it is typically high in BOD, phosphorus, and nitrogen. This method eliminates the need for dewatering and minimizes any potential air pollution problems; on the other hand, maintenance may be significant. Another method currently under study that would also be appropriate on low solids sludges is the atomized suspension technique. The sludge would be thickened to about 8% solids, and then ground to reduce the particle size. The fine sludge would then be atomized and sprayed into the top of a vertical reactor, where combustion would occur. This method would also eliminate the dewatering step, but the process is still under study.[3]

FINAL DISPOSAL

There have recently been many projects devoted to studying the effects of using municipal sludge as a soil conditioner; as a liquid (Figure 4-22), dewatered (Figures 4-23 and 4-24) or dried. Chicago has the largest operation using liquid sludge. The city presently barges the sludge 200 miles to a 10,000-acre previously strip-mined area. The sludge is planned for restoration to productive use. Houston and Milwau-

FIGURE 4-22.   Land spreading of liquid sludge. (*Photo courtesy Wisconsin DNR.*)

FIGURE 4-23. Incinerator settling pond. The particulates from the combustion are frequently collected by wet scrubbers. (*Photo courtesy Wisconsin DNR.*)

FIGURE 4-24. Typical dried water treatment sludge, ready for disposal. (*Photo by David Garot.*)

kee are two major cities that market dried sludge—Houston's being sold to a Florida contractor for use in citrus groves, and Milwaukee's being the very familiar "Milorganite," available in many stores across the country.

Landfilling is the most common method used for sludge disposal. Usually, before the sludge can be landfilled, it does require digestion to avoid odors, insects, and water pollution. All types of dewatered sludges can be disposed of by landfill, as can many other types of solid wastes. The sludges can be transported to the landfill site by truck, train, pipeline, or barge (Figure 4-25).

Ocean disposal is still a possibility for some coastal cities. The EPA has extended the deadline on its ban from (originally) 1981, in spite of serious environmental concerns.

However, there are definite criteria that must be met before the EPA will issue a permit. These include the following restrictions:

1. *Absolutely prohibited substances:*
   Biological, chemical, or radiological warfare agents
   High-level radioactive wastes
   Inert floating materials.
2. *Materials prohibited except in trace amounts:*
   Mercury
   Cadmium
   Organohalogens
   Oils and greases

FIGURE 4-25.   Preparing the sludge for landfill. (*Photo courtesy John Kirchner and the Green Bay Metropolitan Sewerage District.*)

3. *Regulated materials, requiring special care:*
   Compounds of arsenic, beryllium, chromium, copper, lead, nickel, selenium, vanadium, and zinc.
   Organosilicon materials.
   Inorganic processing wastes such as cyanides, chlorides, fluorides, and titanium dioxide.
   Petrochemicals such as aliphatic solvents, amines, phenols, detergents, plastics, and aromatics.
   Biocides such as herbicides, insecticides, and pesticides. Biodegradable organic matter. Radioactive wastes not prohibited. Immiscible materials such as gasoline. Hazards to navigation. Acids and alkalies. Containerized wastes. Living organisms.

Many cities, particularly on the East Coast, are in severe difficulties as to available solid waste disposal sites. Thus, though ocean dumping is being phased out as quickly as feasible, some municipalities continue to use our waterways for their wastes.

TERTIARY METHODS

Only a few municipal systems use any form of tertiary water treatment. However, industries that have their own treatment facilities frequently use these more advanced methods, either in conjunction with the conventional primary and/or secondary treatment methods, or to replace the conventional methods. Many industries use tremendous amounts of water, and thus find it very advantageous to recycle much of it. Though secondary treatment usually removes 85% of the BOD and suspended solids, the nitrogen, phosphorus, soluble COD, and heavy metals usually remain in the effluent. The tertiary methods, primarily chemical in nature, can remove the majority of these remaining pollutants, and make the water suitable for in-plant purposes.

Filtration

Various forms of filtration are available to remove a greater percentage of the suspended solids than can be done by conventional methods. Microscreening, the passing of the effluent through a fine (23-μm) metallic filter fabric, is one type of filtering. The fabric makes up the surface of a horizontal rotating drum. The water flows in one end of the drum and then out through the fabric sides. The solids are retained inside, and can be flushed to a hopper for return to the secondary treatment. This method can remove 89% of the remaining suspended solids, 81% of the BOD, 30% of the total organic carbon, and 76% of the turbidity.

Filter beds composed of very fine particles of sand, coal, and/or garnet are also commonly used. In general, several different substances are mixed, forming a multimedia filter graduated with coarser particles at the top and finer particles near the bottom. Reversing the direction of flow, "backwashing," is necessary to clean the filter. The filtration can be accomplished by only gravitational forces, or pressure can be applied.

In order to remove more of the pollutants, particularly the phosphorus, it is necessary to institute some type of tertiary flocculation before the filtration step.

Adsorption

Dissolved organics often cause tastes, odors, and color in the secondary effluent. They may also be toxic to plant or animal life. Often these organics, called "refractory organics," can be removed by adsorption onto activated carbon. This carbon is usually granular and very porous, and hence has a very large surface area (the surface areas of typical samples are measured in acres). The adsorption is a surface phenomenon; hence, the greater the surface area, the greater the adsorption. After the adsorption capacity of the carbon has been reached, the activity is regenerated by heating the carbon to about 1700°F in air and steam. This burns off the organics, thus preparing fresh surfaces for further adsorption.

Usually, the secondary effluent is passed through tubes filled with activated carbon. It is also possible to use fluidized beds for the carbon; the carbon bed is suspended in a stream of the water to be treated. Powdered carbon is another possibility. The powder is mixed with the waste water, and then, after several minutes, allowed to settle. The powdered form is much more difficult to recover (coagulants are usually necessary to get it out of water solution) and also to regenerate for reuse, though it is less expensive to purchase originally.

Chemical Oxidation

The dissolved organics can also be removed by chemical oxidation. This method can be used either alone or after treatment with activated carbon. Chemical oxidation is stronger than the biological oxidation that occurs in the secondary treatment, and it is likely to force the remainder of the organics to react. Likely oxidants include O_3, hydrogen peroxide (H_2O_2), O_2 with or without catalysts, chlorine or its derivatives, and various oxy-acids.

Electrodialysis

Dissolved inorganics, the mineral content of the water, also must often be removed by tertiary treatment methods. Electrodialysis is one method that can remove these dissolved substances.

When an inorganic salt is dissolved in water solution, it ionizes to produce positively charged cations and negatively charged anions. When an electrical potential is then impressed across the solution, the cations migrate to the negative electrode and the anions to the positive electrode.

Semi-permeable membranes are commercially available that allow the passage of ions of only one sign: cation-exchange membranes are permeable only to positive ions and anion-exchange membranes are permeable only to negative ions. If a series of these membranes is placed alternatively in the solution, and the voltage is applied, the solution between one pair of electrodes becomes clarified, the ions concentrating in the solutions in the adjacent compartments (Figure 4-26).

The ion-exchange membranes usually look like a sheet of plastic. Cation membranes generally consist of crosslinked polystyrene.

$$(R - CH_2 - CH - CH_2 - CH - CH_2 - R),$$

to which sulfonate groups ($-SO_3H$) have been added. The sulfonate groups ionize in water solution, releasing a hydrogen ion (H^+). The anion membranes, on the other hand, have quaternary ammonium groups ($-NR_3OH$) attached to polysty-

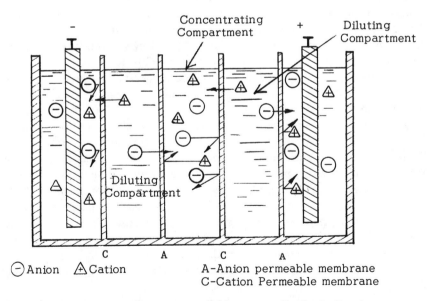

FIGURE 4-26.   Electrodialysis. An electric current concentrates the dissolved ions in compartments adjacent to those between the electrodes.

rene. The OH^- ions can be released in water solution. These mobile H^+ or OH^- ions can be exchanged with similarly charged ions in the contacting solutions, effectively allowing the other ions to "pass through" the membrane. If both processes are carried out on the same solution, the H^+ and OH^- released can then react, forming water.

Electrodialysis has been used for many years for the desalination of brackish water. Since 1954, it has been used for industrial applications. The water can usually be desalinated by this means to between 100 and 200 ppm total dissolved solids. New modifications recently have resulted in as low as 3 ppm total dissolved solids in specific cases. If this purity is not adequate, the electrodialysis can be followed by ion-exchange procedures. Typical applications are in petroleum refining, pulping, petrochemical production, mining, glass etching, and automobile manufacturing. Whey desalination is a major application of electrodialysis. Japan has used electrodialysis as the first step in recovering sodium chloride (table salt) from sea water[4].

Ion Exchange

Ion exchange is the basis for the familiar home water-softening techniques. Industrially, it can be used to remove undesirable dissolved inorganic salts from the water, sometimes simultaneously recovering useful process byproducts.

The basis for the ion-exchange procedure is an ion-exchange resin. There are a number of suitable resins available, including the naturally occurring zeolites and many synthetic resins. These resins can exchange their positive or negative ions for other positive or negative ions, which are dissolved in the water sample. The resins are generally one of two types: either they exchange their hydrogen ions for dissolved cations (metal ions), or they exchange their hydroxide ions for dissolved anions, such as chlorine ions (CI^-). If a particular water sample flows through both types of resins, the resulting "dissolved" ions will be H^+ and OH^-, which would then react to form more water molecules.

These resins can be regenerated for reuse. Cation-exchange resins are regenerated with strong acids; anion-exchange resins are regenerated with strong bases. These resins actually consist of large macro-ions. These macro-ions, to be suitable for ion-exchange resins, must have a greater attraction for the dissolved ions than for the associated H^+ or OH^- ions. For example, a cation-exchange resin, $Resin^- \cdot H^+$, readily replaces the H^+ with sodium ion, Na^+, forming $resin^- \cdot Na^+$ and releasing H^+ into the water. A similar situation exists for anion exchange resins. These resins can be regenerated, however, by a concentration effect: even though the bonding is greater between the resin and, say, the Na^+, the Na^+ can be replaced on the resin if the H^+ concentration is great enough in a solution that flows through.

These ion-exchange methods can be used for phosphates and nitrogen compounds, which is often very important. Usually, they are more efficient than is really needed. Organics can foul the resins. It is difficult to dispose of the regenerating wastes, and their operation is costly. Attempts have been made to mix water thus treated with untreated water, decreasing the overall quality, but also reducing costs.

Reverse Osmosis

Another treatment process, for both dissolved organic and dissolved inorganic materials, is reverse osmosis. This is a membrane process as is electrodialysis, but it operates on a different principle.

Normally, when two solutions of varying concentrations are separated by a semi-permeable membrane, the water migrates into the solution of higher concentration, equalizing as much as possible the concentrations. The difference between the two concentrations provides the driving force for this behavior. The osmotic pressure, the amount of pressure required to be applied to the more concentrated solution to stop the water migration, is thus a measure of this driving force.

The direction of water flow can, in fact, be reversed by applying pressure greater than the osmotic pressure to the more concentrated solution. The water would then flow from the side with a greater concentration of impurities to the side of the membrane with a lower concentration. In this way, one side of the membrane would become very concentrated in the contaminating dissolved organic and inorganic impurities, and the other would have essentially pure water.

The membrane itself is typically a cellulose acetate type, porous except for a dense 1-μm thick layer on one side. This type of membrane will reject the flow of dissolved contaminants, though the mechanism for this behavior is not known. Tests made on an early installation in Pomona, California showed that 80–85% of the feed water could be recovered and the reduction in dissolved contaminants was 88% for total dissolved solids, 84% for COD, 98.2% for phosphate, 82% for NH_3, and 67% for nitrates. This reduction was on water that had previously been treated by the normal primary and secondary methods, and also by carbon adsorption.[5]

Phosphorus and Nitrogen Removal

Because of their nutrient possibilities, it is particularly important to remove as much as possible of the phosphorus and nitrogen from the effluent water.

Phosphorus can usually be removed by only minor additions to the normal primary and secondary treatment methods—these additions being precipitating agents. Usually, the phosphorus is present in the effluent from the secondary process as orthophosphate ions (PO_4^{3-}); some is also present in organic compounds. If precipitating agents such as aluminum sulfate (alum, $Al_2(SO_3)_3$), lime (CaO), or ferric chloride ($FeCl_3$) are added, they precipitate the phosphates as fairly complex inorganic compounds. The exact products depend on the conditions. For example, in neutral or slightly acidic solutions, aluminum phosphate and aluminum hydroxy-phosphates are formed; at alkaline pH values, calcium phosphates and hydroxapatites precipitate. These precipitates may settle, be filtered, and/or be further flocculated. Actually, these precipitating chemicals can be added at various stages in the treatment: before the primary settling, in the aeration basin of an activated sludge process, or after the secondary settling. If, for example, sodium aluminate is added to the aeration basins, the phosphorus-containing precipitates become intimately associated with the activated sludge, and can settle out with it in the final settling tank.

The biological treatment processes themselves actually remove some of the phosphorus. This is attributed to the fact that many microorganisms take up more phosphorus than they need to grow.

As discussed above, ion exchange and reverse osmosis are also effective in phosphate removal. A method of alumina adsorption comparable to carbon adsorption has also shown potential for phosphate removal.

Nitrogen removal is more complex to consider. After secondary treatment, most of the nitrogen is present as NH_3. This can be treated either biologically by a nitrification-denitrification procedure, or by physical-chemical methods such as ammonia stripping, ion exchange, or excess chlorination.

Nitrification-denitrification is the conversion of the NH_3 to nitrates by nitrifying bacteria, followed by the conversion of the nitrates to nitrogen gas by denitrifying bacteria. Nitrification requires O_2, which must be supplied to the process. This step can be carried out in a system similar to a trickling filter, located after the secondary treatment. The process is slow, so adequate time must be provided. If the effluent is to be discharged to a swiftly moving body of water, it may be desirable not to carry the treatment any further, for the nitrates do provide certain advantages, such as supplying oxygen to sludge beds and thus reducing the BOD.

If, however, the water is to be discharged to a slower moving body of water, the nitrates must be removed because of their nutrient potential for algae. The denitrification must be done anaerobically, either in an activated sludge type system or a columnar system. If the bacteria are added to the waste water and fed methanol as a nutrient, they decompose the nitrates in order to obtain the oxygen needed for growth. The process then releases harmless (N_2) to the atmosphere.

This process, though it creates no additional pollutants or sludge, does require a fair amount of space and significant energy, and the loss of microbes due to an upset by toxic materials can result in serious disruptions.

Gaseous NH_3 can also be directly "stripped" from the water by using air. In the water, the following equilibrium exists:

$$NH_3 + H_2O \rightleftharpoons NH_4^+ + OH^-.$$

If the pH is raised to about 11.5 (the solution made more basic) by adding lime (this simultaneously also precipitates the phosphates), the equilibrium shifts to the left. If the solution is then agitated vigorously, the NH_3 will be driven from the system. Usually, the NH_3 is in low enough concentrations so that it does not create air pollution problems.

This process is simple, reliable, and economical. Cold weather does, however, decrease the efficiency, and freezing weather may make the process inoperable. Energy requirements are comparable to the biological method.

Selective ion exchange is possible, using in particular a clinoptilolite exchange resin, a form of zeolite. Regeneration is done using concentrated salt solutions. This method is very efficient but complex, and capital costs are high.

Breakpoint chlorination is the process of adding so much chlorine that the NH_4^+ reacts with it, simultaneously releasing N_2 gas. This is a highly efficient process with low capital costs and no sludge disposal problems, but the effluent is highly chlorinated, which may present problems.

When discussing industrial water treatment, it is difficult to generalize, for the sources as well as the appropriate treatment methods are very diverse. Often, even one industry will use a combination of waste disposal methods, including conventional primary and secondary treatments, incineration, and deep well injection. Frequently, process modification may be the best and most economical method of handling wastes.

References

1. *Environmental Pollution Control Alternatives: Municipal Wastewater.* U.S. EPA Technology Transfer, EPA-625/5-76-012.
2. Trump, John C., "Disinfection of Municipal Sludges by High Energy Elections." *Pollution Engineering*, September 1978, p. 49.
3. *Cleaning Our Environment—The Chemical Basis for Action.* American Chemical Society, Washington D.C., 1969, p. 119.
4. Leitz, Frank B., "Electrodialysis for Industrial Water Cleanup." *Environmental Science and Technology* 11, *No.* 2, February 1976, p. 137.
5. *Cleaning Our Environment.* p. 130.

5

Solid and Hazardous Waste Disposal

SOLID WASTE

When one thinks of solid waste, often the problems and processes that immediately come to mind are those associated with municipal disposal (Figure 5-1). These substances generally are, of course, the products of various industries, but their disposal is not directly the responsibility of the industry that created them. The industries have their own problems—their own types of solid wastes that must be disposed of.

The ideal solution, economically, energetically, and environmentally, would be to recover and reuse many of the solid wastes. Many industries have been attempting to recover their wastes, with varying degrees of success. As with most industry-related issues, pollution control and economics are inseparable. The primary responsibility of an industry official is to protect his or her company's financial position; if not, the company will soon be out of business, and the shareholders will suffer. The financial incentive may be to avoid fines, court cases, or costly enforcement squabbles, or it may be byproduct recovery, but, unless the incentive is there, little progress will be made.

There are several types of solid wastes an industry may have to handle. There are, of course, the sludges that result from water treatment and that have already been discussed. There are also the process solids, such as collected particulates and slags. Many of these are composed of various minerals, though their form and actual chemical composition may vary significantly, depending upon the source.

More than one billion tons of solid wastes are produced annually by the minerals processing industries alone. These ores usually contain only small percentages of the desired substances (such as copper, iron, gold, or silver); thus, the spent ores, or tailings, accumulate very rapidly. Tailings are typically composed of silica

FIGURE 5-1.   Burning of an open dump. (*Photo courtesy Wisconsin DNR.*)

(sand), and various silicates and carbonates of calcium, magnesium, and possibly aluminum. These tailings, often consisting of very fine particles, are piled near the processing plant, creating a nuisance because of their size and physical instability; plant growth often must be encouraged to stabilize the piles. Few recovery methods have been found to be economically feasible for many of these wastes.

Many industries also generate fly ash, the coal ash that results from, among other things, power generation. Fly ash is one substance on which much research has been done, looking for more and better ways for its utilization.

SANITARY LANDFILLS

If the various wastes must be landfilled, they must usually be disposed of in "sanitary landfills." A sanitary landfill is one in which the wastes are deposited, spread into layers, compacted, and covered daily with earth (Figures 5-2 through 5-6). There are two basic methods by which to do this: area and trench. Area

FIGURE 5-2. This landfill cell is just starting to be filled. The waste is spread on a surface layer of very fine sand.

FIGURE 5-3. Spreading the cover material. (*Photo courtesy Wisconsin DNR.*)

FIGURE 5-4. Covering almost completed. (*Photo courtesy Wisconsin DNR.*)

FIGURE 5-5. Moving more cover material close to another fill area. (*Photo courtesy Wisconsin DNR.*)

FIGURE 5-6.   Almost filled site. (*Photo courtesy Wisconsin DNR.*)

landfills are suitable where land depressions exist, such as canyons, ravines, or valleys. The waste is simply spread, compacted, covered, and then compacted again until the area is filled, after which a thicker layer of earth is spread over the whole area. In a trench landfill, the trench is first cut into the ground, and then the spreading-compacting sequence is carried on as before. This method is suitable for flat or slightly rolling land, if the water table is low enough and the soil is deep enough for trenching.

Sanitary landfills are particularly important in disposing of biodegradable wastes where harmful gases and microbes may form; often they are not used for mineral wastes such as cement dust. The anaerobic decomposition of organic materials leads to the production of gases such as methane (CH_4) and hydrogen sulfide (H_2S), as well as CO_2 and N_2. CH_4 can be an explosion hazard if high concentrations accumulate in enclosed areas. H_2S smells similar to rotten eggs, and is toxic in higher concentrations. CO_2 can dissolve in ground water, forming carbonic acid, which will dissolve the rocks and mineralize the water. To minimize the effects, the gases can be vented from deep in the landfill to the atmosphere. This will reduce the possible hazards of CH_4 and H_2S by not permitting any buildup of the gases.

Another problem associated with landfill sites is leaching (Figures 5-7 and 5-8). Ground water or surface water that infiltrates the solid waste can dissolve some of the waste substances or form suspensions of solid matter and microbial waste products. This leachate can then percolate through the soil to either the underground water systems or to nearby lakes and streams. This problem is not confined to

FIGURE 5-7. Leachate from old landfill site. (*Photo courtesy Wisconsin DNR.*)

FIGURE 5-8. Leachate from old landfill site. (*Photo courtesy Wisconsin DNR.*)

biodegradable wastes; mineral wastes can also have major leaching problems, the leachate often being either acidic or basic, depending upon the source.

Leaching can be minimized by lining the site with fine-grained soils, clays, or synthetic liners; locating the site in areas far from natural water systems; and covering the site with properly graded, nearly impervious soils such as clays. Other areas must have sumps to collect the leachate (Figure 5-9) or special drainage systems (Figure 5-10).

Another major problem associated with landfill sites is simply the scarcity, in many areas, of suitable sites located reasonable distances from the various plants. As more industries dispose of their wastes by landfill, the number of suitable sites gets smaller and smaller, and the costs for disposal rise. Some types of waste materials will eventually decompose, providing areas that can be reclaimed for recreational areas; other types of wastes, particularly mineral wastes, do not generally produce such amenable areas. Many such areas are difficult to reclaim, particularly if future construction is desired. Recovery of the waste materials is currently the most economically favorable method of solid waste handling for many industries, such as the cement industry,[1] and the probability is that the feasibility will increase as shortages loom for the U.S. in nonfuel mineral supplies such as chromium, cobalt, manganese, titanium, biobrium, aluminum, tin, strontium, sheet mica, and similar materials. According to Philip M. Smith, Assistant

FIGURE 5-9.   Sump located on landfill site. (*Photo courtesy Wisconsin DNR.*)

FIGURE 5-10.   Leachate collection. (*Photo courtesy Wisconsin DNR.*)

Director of the Office of Science and Technology, "In some cases known reserves will be exhausted in 10 to 20 years.[2]

RESOURCE RECOVERY

Much interest in recent years has focused on recovering industrial wastes rather than disposing of them by landfill. The following is a brief summary of some uses of large-scale industrial wastes.

Coal Ash

The burning of coal produces both bottom ash and fly ash. Bottom ash remains in the bottom of the boiler after burning; fly ash is entrained in the flue gas. Bottom ash is a coarse, gritty material in contrast to very fine fly ash.

In the United States, approximately 550 million metric tons of coal are burned annually. On the average, 11% of this is ash. Approximately 8-9% of this total ash is bottom ash, and the remainder is fly ash. Considerable efforts have gone into the study of fly ash as a soil amendment.[3] A limiting factor appears to be the boron content.

Fly ash has been used fairly widely as an additive for cement. The fine fly ash can be added to the ground cement clinker, increasing for some purposes the desirable cement characteristics. For example, the U.S. Corps of Engineers uses fly ash in much of its concrete. Unfortunately, this is not the solution for fly ash disposal. Many cement companies have found the cement not to be marketable, primarily due to its dark color.

Fly ash is also used as one of the raw materials for the production of sintered lightweight aggregate, such as is used in concrete blocks and other precast forms. It can be a raw material in bricks. Recent studies have shown its feasibility as a plastics filler.[4]

Most of these uses, however, even if fully implemented, will never generate sizable markets for fly ash. More large-scale uses are needed. Some promising applications that have been implemented are as a filler in asphalt pavings, a soil stabilizer and/or fill for embankments, and in the bases for road beds. However, additional innovative applications are necessary to utilize the vast amount of fly ash generated daily.

Bottom ash, in contrast to fly ash, shows significant potential as an agronomic soil additive that will not be detrimental to the soil, crops, or the environment.[5] Bottom ash is a material about the size of sand or fine gravel. Because of its coarse size, it is suitable for mixing with clayey soils to modify the texture. It is also a minor source of nutrients: because it is a fused silicate, the rate of nutrient or other element release is very slow. Thus, it does not exhibit the potential toxic characteristics of some fly ashes.

Waste Treatment Sludges

The chemical content of waste treatment sludges varies dramatically with the type of process industry. Sludges from metals finishing companies, for example, are high in heavy metals and are thus classed as hazardous. Paper mill sludges, on the other hand, are primarily wood fibers, dead microorganisms from secondary treatment, and perhaps papermaking fillers such as clay. These latter sludges can be recycled fairly easily. One of the primary uses is as a fuel.[6]

Pulp and paper companies produce massive quantities of sludge in their waste treatment operations. A medium-sized mill, for example, might generate 40–50 dry tons of sludge per day. This sludge is typically produced at 1.5–3% solids. Screw presses can further dewater it to, perhaps, 45–50% solids. One way to treat this material, with about half the heat content of coal, is to burn it in a hog fuel boiler, to produce process steam. The resulting ash, typically 20% of the original sludge by weight, is significantly less of a disposal problem than was the original sludge.

HAZARDOUS WASTE

Hazardous wastes are those that are ignitable, corrosive, chemically reactive, or toxic.[7] Many wastes from industrial plants are toxic, and, once collected, the plant is faced with the question of treatment and final disposal. In the past, such

FIGURE 5-11.   Even after over two decades of concern regarding the proper disposal of hazardous wastes, barrels such as these can be found in wooded areas "off-the-beaten-track." (*Photo courtesy J. V. Manufacturing Co., Inc.*)

substances were often simply drummed and then disposed of in a landfill area. This was generally unsatisfactory.

The RCRA land ban restrictions (Chapter 2) have totally changed these practices. All hazardous wastes now must be treated prior to disposal to meet prescribed contaminant levels.

TREATMENT

There are a number of possible treatment procedures. Three common methods are incineration, biological treatment, and stabilization/solidification.

Incineration[8]

A prevalent method for disposing of organic wastes is by incineration. If a waste is burned, the only potential problem one might have in the future is with the ash; if the waste is buried, it may instead become a future Superfund site.

Customarily, incinerators are used for concentrated materials with high organic content. However, inorganics that have been contaminated with only parts per million of toxic chemicals such as polychlorinated biphenyls can also be incinerated. The high temperatures drive the toxic materials from the inorganic and then oxidize them, leaving the inorganic residual.

Incinerators have been designed to handle waste gases, liquid wastes, and solids and combination wastes. Waste gases can be burned by systems as simple as a flare. For systems with too low a heat content for self-sustained combustion, thermal oxidation (using a supplementary fuel) or catalytic oxidation can be employed.

Liquid wastes encompass materials such as contaminated solvents and process residuals (still bottoms or sludges). These materials can be burned in a liquid incinerator, so long as the waste can be pumped and then atomized for combustion. Typical liquid incinerators operate at 1000–2400°F with residence times of the waste of 1–3 seconds or perhaps longer. Inclusion of heat recovery units can make the systems more economically feasible.[9]

Solids and combination wastes are more problematical to handle. By definition, they encompass all materials, including sludges, that are not pumpable and atomizable. A variety of different incinerator designs are available, including two-chamber controlled air incinerators, rotary kiln incinerators with afterburners, and fluidized-bed incinerators. Commonly, baghouses or electrostatic precipitators have to be used for particulate collection.

Plasma Arc Furnaces[10]

Long used in steelmaking, plasma arc furnaces recently have seen some application in hazardous waste incineration. Their high electrical power demands have limited their use, but for some wastes, no other type of incinerator will suffice.

Plasmas are high-temperature, overall electrically neutral, gaseous streams, consisting of positively charged atoms and molecules and the electrons that were stripped from them. The pressure of the plasma determines its temperature and transport properties, permitting the destructive capacity to be varied to meet the specific needs of different waste materials.

The temperatures of plasmas commonly reach 10,000°C. At these extremely high temperatures, the hazardous wastes are decomposed within milliseconds, without the generation of secondary combustion products, thus simplifying flue gas cleanup.

Wastes that have been treated by plasma arc furnaces include mixtures varying from 100% metal to 100% clay (making it ideal for treating problems such as buried waste drums), organic compounds such as PCBs, toxic flyashes, and mixed hazardous and radioactive wastes. The organic components are dissociated into their elements, which then recombine to form carbon monoxide, hydrogen, water, and (if the waste is chlorinated) hydrochloric acid. Heavy metals are immobilized in a vitrified, glassy, nonleachable slag that can pass the EPA Toxicity Characteristic Leaching Procedure (TCLP) tests.

Supercritical Hazardous Waste Decomposition[11]
The use of "supercritical water" has shown potential as a promising new method for the destruction of hazardous wastes. PCBs, explosives, sewage sludge, and pulp mill wastes are among the materials that have been decomposed by this technique.

The concept of a supercritical fluid dates back to the early 19th century. The decomposition of organics in supercritical water was first observed in the mid-1970s. For the next decade, much of the developmental work was unpublished. It wasn't until September 1990 that the North Atlantic Treaty Organization sponsored a workshop that focused on ways to develop a technology for destroying toxic molecules in supercritical water.

As the temperature and pressure of a liquid and gas in equilibrium increase, the densities of the two phases become indistinguishable. Above the "critical point," the two phases are identical, and the substance can be described only as a fluid.

The conditions of interest for this particular application are above the critical temperature (374°C for water) and near the critical pressure (218 atmospheres, or 2.21×10^7 N/m^2 for water). Under these conditions, the properties of the water are intermediate between those of typical gases and liquids. Nonpolar organic wastes, such as PCBs, are soluble in all proportions and, in the presence of an oxidizer such as oxygen or hydrogen peroxide, can react to form water, carbon dioxide, and (for halogenated organics) simple acids. Greater than 99.9% of the hazardous materials can be detoxified in just a few minutes, leaving an innocuous aqueous residual that in many cases can be disposed of without further treatment.

A supercritical reactor operates as a closed system, unlike an incinerator, thus it emits no hazardous materials into the atmosphere. Operating temperatures of 500–600°C are well below the 2000–3000°C often found in incinerators, resulting in fewer NO_x emissions. The lower temperatures also lead to a lower supplementary fuel requirement: an ordinary incinerator requires a waste that is about 30% carbon to be energy self-sufficient; supercritical reactors can be self-sufficient if the wastes contain as little as 10% carbon.

There are, however, a number of problems that must be solved before there is large-scale use of the technology. Supercritical water is very corrosive, especially when used to treat halogenated compounds, potentially requiring construction using special alloys. Not only could corrosion destroy the main body of the reactor, but corrosion products could plug inlet and outlet tubes. Concern about operating under extreme conditions of temperature and especially pressure has also slowed full-scale development and implementation of the process.

Stabilization/Solidification[12,13]

If a waste is to be stabilized (that is, converted to a form that will resist leaching) and then solidified prior to disposal, the treated product must:

1. Withstand a a pressure of 50 psi when applied in an Unconfined Compressive Strength test; and
2. Meet certain maximum leachate concentrations when subject to the Toxicity Characteristic Leaching Procedure (TCLP) determination.

These analytical tests were developed to approximate conditions at the bottom of a landfill site.

The EPA has specified stabilization and solidification technology as the best demonstrated and available technology for the disposal of heavy metals. For most metals, the technique provides excellent results for long-term immobilization. It is, however, necessary to choose the proper combination of stabilizing/solidifying agents, since many metals can be quite soluble and hence subject to leaching, in both very acidic and very alkaline environments.

Stabilization of waste can include a variety of chemical processes. pH adjustment is often critical, since solidification processes are frequently pH-sensitive.

FIGURE 5-12.   Hazardous wastes must be able to undergo a 50 psi unconfined compressive strength test, as illustrated, as well as meet certain maximum leachate concentrations when subjected to a TCLP procedure, before they can be safely disposed.

Acids and bases are generally neutralized; heavy metals are brought to a pH of 8–10, to encourage precipitation as their hydroxides. The wastes can be solidified using two general types of products: sorbents and encapsulating agents.

Sorption includes both absorption and adsorption. Absorption is the uptake of a substance into the bulk sorbent via the pore structure, by diffusion through the solid portion of the sorbent, or by vapor diffusion through the empty space in the material. Adsorption, in contrast, is a surface phenomenon. Liquid or gaseous molecules are attracted to and retained on the sorbent surface by chemical bonding or physical forces. Both of these processes occur simultaneously, and thus are jointly referred to as sorption.

Sorbents are marketed in an assortment of types, including inorganic solids (for example, silicates), polyolefins, and cellulosic types. Only the silicate types are likely to meet EPA guidelines; the polyolefins and cellulosic sorbents should be used solely for emergency spill control, such as off-shore oil spills.

Silicate-based sorbents are available from a diversity of sources, including vegetable products and volcanic ash. The silicate ions (SiO_3^-) can bond together to form either a cross-bonded sheet, as found in clays, or a polymeric chain. Smaller SiO_3^- units occur in felspars (for example, $KALSi_3O_5$) and geolites (for example, $NaAlS_2O_6 \cdot H_2O$). Finely divided silicates have a high porosity, ensuring significant liquid and vapor transport across and within particles. Most are quite stable and relatively nonreacting with all organic and inorganic chemicals except HF.

An untreated clay is hydrophilic. Thus, though it can chemically react with ("fix") an assortment of inorganic contaminants, it generally is useless for organic pollutants. However, clays can be modified by the ion exchange process with selected organic compounds that have a charged site, rendering the clay/organo complex hydrophobic.[14] Often the organic molecule used is a quaternary amine, an ammonium salt substituted with four long-chain carbon groups. The quaternary amine exchanges with sodium, calcium, and magnesium ions on the clay surface, permitting the organically modified clay (organoclay) to then sorb a variety of organic compounds.

Encapsulation methods can be of two types: macro and micro. Macro encapsulation is the encapsulation of a large quantity of waste; that is, an entire waste container. The process consists of two essential elements: a stiff, weight-supporting component and a tough but flexible seam-free jacket.

Microencapsulation is the encapsulation of pollutant material on the molecular level. Most of the microencapsulation techniques are founded on the reduction of the surface-to-volume ratio by the formation of a hard, monolithic mass with a low permeability. Many of the curing reactions involve hydration, further stabilizing the waste by incorporating any free water into their structure. Microencapsulating agents include thermoplastic materials (asphalt, tar, polyolefins, and epoxies), organic polymer resins (urea-formaldehyde, polyacrylates, and polyacrylamides), portland cements, and lime/fly ash mixtures.

Sorption and encapsulation techniques can be used alone or in conjunction with each other. Portland cement or lime/fly ash mixtures are commonly considered for solidification of heavy metals. They generally are not satisfactory for organics when used alone, due to leaching.

Biological Treatment

Conventional biological treatment using activated sludge, trickling filters, waste stabilization ponds, and/or aerated lagoons are another option for some wastes. However, not all wastes can be detoxified or decomposed into a nonhazardous form. Moreover, wastes considered toxic might also be toxic to the microorganisms.

DISPOSAL

Once the wastes are treated, the problem of disposing of the residual wastes often remains. Two disposal techniques are common: hazardous waste landfills and underground injection.

Hazardous Waste Landfills

Secured landfills, those that can produce no toxic leachate, are available in some locations and can sometimes be used to dispose of toxic wastes. These landfills are located in thick, natural clay deposits, or employ specially designed liners and leachate collection systems. Often, however, appropriate sites are too far from the plant that generates the waste to make them an economically viable disposal method. In addition, the tipping fees (the charge per ton to dispose of materials) are very high, the number of available hazardous waste landfills is small, and it is very difficult politically to site new disposal facilities.

Underground Injection[15]

Underground injection, sometimes called underground sequestering, is the controlled emplacement of fluids into selected, deeply buried geologic formations through specially designed and monitored wells. The method was developed and used by the gas and oil industry in the 1930s for disposal of brines, and applied to the disposal of industrial wastes in the early 1950s. As regulations regarding surface disposal of wastes became more stringent, underground injection gradually increased.

The wastes that can be disposed in this manner can be either hazardous or nonhazardous. Acids (HCl, H_2SO_4, and HF), caustic soda (NaOH), pesticides, fluoride, mercury, arsenic, vanadium and chromium compounds, and chlorinated hydrocarbons are some of the more common wastes disposed of in this manner.

The feasibility of a deep well injection system depends on the geologic environment. At each well a permeable injection zone to hold the waste is sandwiched between thick, impermeable rock layers above and below to confine the wastes. The injection zone may be of, for example, porous and permeable dolomite or limestone, or of sandstone. The wastes are always injected below all underground sources of drinking water; hence, well depths may be 1500–6000 ft.

Other Options

A company that generates hazardous waste has another option: it can hire one of the types of firms that will remove hazardous wastes or contaminated fluids from its site.[16] These firms consist of transporters, which pick up and haul away wastes; transfer and storage companies (waste brokers), which collect wastes from several firms and combine them into shipments extensive enough to be taken to a large disposal firm; and treatment, storage, and disposal firms, which will dispose of, recycle, and/or treat wastes.

This option does not relieve the generator of any responsibility for the wastes. Therefore, if a firm chooses to entrust this responsibility to a second company, it is necessary that the disposal firm be selected and evaluated exceedingly carefully.

BIOREMEDIATION

Unintentional contamination of soils by toxic compounds is a continuing problem. For example, the U.S. General Accounting Office has estimated that over half of the approximately 5000 hazardous waste landfills administered by the EPA are leaching dangerous chemicals.[17] Transportation accidents persist: semitrailers overturn, and trains derail, spewing forth toxic substances. Oil spills continue, such as illustrated by the grounding of the Exxon Valdez in Alaska's Prince William Sound and the releases of petroleum into the Persian Gulf during the Iraqi occupation of Kuwait. Site remediation is, therefore, of perennial importance. Biodegradation techniques are one of a number of methods that are currently under study.

Biodegradation is the simplification of the structure of an organic compound by the breaking of its intramolecular bonds by microbiological, usually enzymatic, catalysis. The simplification can be subtle, such as modifying only a substituent functional group, or it can be complete mineralization—that is, conversion into its inorganic constituents (CO_2, NO_3^-, SO_4^{2-}, PO_4^{3-}). Bioremediation is the application of this microbiological catalysis to pollutant compounds, to eliminate environmental contamination.

Most recent interest has focused on *in situ* bioremediation studies. *In situ* is Latin for "in its original place." *In situ* treatments thus focus on destroying or otherwise immobilizing pollutants where they are found in the landscape. Removing the contaminants from the spill site (often along with the soil in the area) for subsequent physical, chemical, or biological treatment in, for example, an incinerator or a bioreactor, is not *in situ* treatment.

The degradation of naturally occurring organic compounds is of critical importance for the successful functioning of the ecosystem. Since the beginning of time, microorganisms have enzymatically decomposed plant and animal matter, completing the carbon cycle that was started by photosynthesis. Bioremediation strategies for pollution control are thus simply applying the innate biodegradation capabilities of microorganisms to a number of "undesirable" organic compounds. Application of the technique can be as simple as applying an appropriate fertilizer to the contaminated area.

Often, it is difficult to verify the effectiveness of a biomediation treatment on a spill; that is, whether or not a spill has actually undergone biodegradation. For example, in recent attempts to apply in situ bioremediation to remove crude oil from Prince William Sound, a visually dramatic loss of dark surface material was observed only on sections of rocky beaches on which an oleophilic fertilizer was spread. However, initially scientists were not certain whether the fertilizer acted as a nutrient source, stimulating microbial activity, or merely as a solvent, mobilizing the crude oil. Further chemical analyses did, in this case, verify that biodegradation did occur.[18]

Though *in situ* biodegradation does appear to be a promising treatment process, further tests are needed to maximize the degradation rates, expand the capabilities to include treatment of complex mixtures of organic compounds, and verify sufficient cleanup to meet the allowable amounts of residual chemicals that have been established by regulatory agencies.

Studies as to the possibility of using naturally occurring organisms to break down specific hazardous chemicals in industrial wastes are also ongoing. For example, Celgene Corp. has developed a process to degrade methylene chloride, a suspected carcinogen.[19] Methylene chloride is used as a solvent in the production of thermoplastic resins, photographic film, paint removers, and pharmaceuticals.

In the Celgene process, effluent containing methylene chloride is treated with nutrient chemicals and sodium hydroxide, and then fed into a proprietary bioreactor containing immobilized microbes. The microbes convert the methylene chloride into water, carbon dioxide, and sodium chloride. The residual chemical is in the parts-per-billion range, well below what can be achieved by conventional distillation.

References.
  1. Sell, N. J. and Fischbach, F. A., "The Economic and Energy Costs of Dust Handling in the Cement Industry." *Resource Recovery and Conservation* 3, *No.* 4, 1979, p. 468.
  2. Chemical and Engineering News, March 20, 1978, p. 16.

3. Adriano, D. C., Page, A. L., Elseewi, A. A., Chang, A. C., and Straughan, I. "Utilization and Disposal of Fly and Other Coal Residues in Terrestrial Ecosystems: A Review." *J. Environ. Quality* **9**, 1980, pp. 333-343.

4. "Fly Ash Shows Promise as Plastics Filler." *Chemical and Engineering News*, May 8, 1978, p. 29.

5. Sell, N. J., McIntosh, T. H., Severance, C. W., and Peterson, A. "The Agronomic Landspreading of Coal Bottom Ash: Using a Regulated Solid Waste as a Resource." *Resources, Conservation and Recycling*, February 1989, pp. 119-129.

6. Sell, N. J., McIntosh, T. H., Jayne, T., Rehfeldt, T. and Doshi, M. "Burning Bulk Screw Press Dewatered and Briquetted Sludge in a Hog Fuel Boiler," *TAPPI Journal* **73 No. 11**, November 1990, pp. 181-188.

7. "Hazardous Waste Disposal." *Science News* **114**, *No. 26*, December 23 and December 30, 1978, p. 440.

8. Brunner, Calvin R., "Incineration: Today's Hot Option for Waste Disposal." *Chemical Engineering* **94**, *No. 14*, Oct 12, 1987, pp. 96-106.

9. "The Fire Next Time," *Environmental Science and Technology* **12**, *No. 2*, February 1978, p. 134.

10. Ondrey, G. and Fouhy, K., "Plasma Arcs Sputter New Waste Treatment." *Chemical Engineering* **98**, *No. 12*, December 1991, pp. 32-35.

11. Shaw, R. W., Brill, T. B., Clifford, A. A., Eckert, C. A., and Franck, E. U., "Supercritical Water: A Medium for Chemistry." *Chemical and Engineering News* **69**, *No. 51*, December 23, 1991, pp. 26-39.

12. Conner, Jesse R., "Fixation and Solidification of Wastes." *Chemical Engineering* **93**, *No. 21*, Nov 10, 1986, pp. 79-85.

13. Sell, Nancy J., "Solidifiers for Hazardous Waste Disposal." *Pollution Engineering*, **XX**, *No. 8*, August 1988, pp. 44-49.

14. Sell, Nancy J., Revall, Mark A., Bentley, William and McIntosh, Thomas H., "Solidification and Stabilization of Phenol and Chlorinated Phenol Contaminated Soils." Stabilization and Solidification of Hazardous, Radioactive, and Mixed Wastes, ASTM STP 1123, Vol. 2, T. M. Gilliam and C. C. Wiles, eds. Philadelphia: ASTM, 1992, pp. 73-85.

15. Brower, Ross D. and Visocky, Adrian P., "Evaluation of Underground Injection of Industrial Wastes in Illinois," Illinois Scientific Surveys Joint Report 2, ENR Contracts AD-94 and UI-8501, 1989.

16. Lucks, John O., "Dispose Hazardous Wastes Safely." *Chemical Engineering* **97**, *No. 3*, March 1990, pp. 141-144.

17. "The Planet Strikes Back." *National Wildlife* **27**, *No. 2*, February-March 1989, pp. 33-40.

18. Madsen, E. L., "Determining in Situ Biodegradation." *ES&T* **25**, *No. 10*, October 1991, pp. 1663-1673.

19. "Microbes Munch on a Toxic Lunch." *Chemical Engineering* **98**, *No. 12*, December 1991, p. 11.

6

Iron and Steel Industries

When considering specific types of industries, one finds that there are often many different production processes, resulting in many different types of pollutants, and also many possible methods of pollution control. Frequently, however, the processes are similar; this allows a discussion of the "general" treatment methods for the "typical" wastes produced. Where feasible, several of the specific processes and the associated pollutants and treatment methods will be considered for each industry.

The iron and steel industry in the United States has undergone radical changes since the 1970s. Once, the steel industry was dominated by corporate giants. Today, however, some 75 companies, many of them quite small, are both competing for U.S. market share and investing in new technologies that are more efficient and pollute less.

The U.S. steel industry capacity has also decreased dramatically since the 1970s, from a high of about 160 million tons/year to a nominal capacity of about 117 million tons/year. A larger fraction of this steel is now produced in what are called "minimills," which produce steel from scrap in electric arc furnaces rather than by the standard basic oxygen process.[1]

The iron and steel industry is very complex and usually vertically integrated. There are four major stages to conventional processing:

1. Mining
2. Iron ore concentration (beneficiation)
3. Blast furnace treatment
4. Steel production.

The first two of these steps are generally done at or near the mining site, whereas the last two processes are done at steel plants located elsewhere.

The pollutants generated are very diverse, as are the processes, and encompass the full range: air and water pollution and solid wastes. Because of the complexity

114

of the steel industry, the processes and associated environmental problems will be considered for each stage separately.

MINING

Iron and steel manufacture starts with the mining operation. Many eons ago, iron oxide and other iron compounds combined with silt under very high pressures to form rock. Now, 20–35% of the iron is found as the oxides hematite (Fe_2O_3) and magnetite (Fe_3O_4) and as carbonates and silicates. This iron ore composes about 5% by weight of the earth's crust. Typical impurities in the iron ore are sulfur, aluminium oxide (Al_2O_3), phosphorus, silica (SiO_2), and titanium. Ore deposits are enriched by weathering. Many of the nonferrous rocks are leached out, leaving an ore such as taconite (Figure 6-1), which is about 28% iron, the rest being silicate rocks and clays.

Mining is done by one of two methods: open pit (strip) or shaft mining. Strip mining is suitable for ores located near the surface, and is used in regions such as around Lake Superior, especially near Hibbing, Minnesota, where the world's largest strip mining operation is found. Deep shaft mines are more expensive and more dangerous to operate, but they do destroy less of the countryside. In the United States mining of one substance or another currently affects approximately 13 million areas, about 0.5% of the total U.S. land area. Of this 13 million acres, 7 million have been undercut, 3 million have been strip mined, and 3 million are

FIGURE 6-1.   Taconite.

serving as sites for waste disposal.[2] Reclamation of strip mined areas is now an important issue. For example, the West Virginia Surface Mining and Reclamation Association recently developed "longwall stripping," a method designed to minimize the environmental impacts, and simplify restoration of the land.[3] From the mine, the ore can be placed directly into trucks or railroad hopper cars to be taken to the ore concentration facilities.

Environmental Problems

The major pollutant associated with the mining operation is the acid mine water. Acid mine drainage pollutes an estimated 10,000 miles of streams in the United States. About 60–70% of the drainage comes from abandoned mines, primarily from underground mines.[4] The mining operation exposes sulfur- and iron-bearing rocks to weathering and erosion. The iron sulfide (FeS) and iron pyrite (FeS_2) ores oxidize to form sulfates, sulfuric acid (H_2SO_4), iron oxides, and ferrous, ferric, aluminium, and manganese salts. Also likely to be present in the drainage water are calcium, magnesium, and sodium salts, CO_2 and silicic acid. The water can come from the normal precipitation, or it can arise from various process waters. The volumes of acid water range from a few hundred thousand gallons per day to tens of millions of gallons per day, depending upon the site. It is estimated that four million tons of H_2SO_4 enter our inland waters annually from this type of source.[5]

One method to reduce the effects of the acid mine water is to minimize its formation at the source. Flooding or sealing the mine to exclude air and thus to prevent oxidation of the pyrites, preventing water from entering the acid-producing areas, or at least minimizing the water's contact time with the reactive surfaces if it does enter, are possible methods.

Once the acid mine water forms, it must be treated by some method. So far, H_2SO_4 recovery has proven to be uneconomical, as has the production of by-products such as metallic ion pigments and rouge; thus, generally, the water is treated and disposed of. The three types of treatment processes possible can be classed as follows:

1. Lime neutralization
2. Limestone neutralization
3. Alternative processes.

The first two of these methods are based on acid-base neutralization, with precipitation of the contaminating metal salts. The lime (CaO) process is the most commonly used at present. In general, this treatment process is not economical, however. Limestone ($CaCO_3$) is cheaper, and very available, and research is progressing on this process. A difficulty is that insoluble ferric hydroxide or sulfate can precipitate on the surface, stopping the reaction. This can be minimized by grinding the $CaCO_3$ before application. The sludge formed by both of these processes, largely hydrated iron oxide, contains high percentages of water. Typically, the sludge is settled in lagoons. Disposal is a difficulty: generally, it has been disposed of in old strip mine cuts, in underground mines, and in new-fill areas.

Alternative methods of treatment, such as distillation, reverse osmosis, and ion exchange, have also been attempted. But except in areas with water shortages, where these methods could be used to produce potable water, they have been uneconomical.

IRON ORE CONCENTRATION

The next step in iron and steel manufacture is the concentration of the iron ore. For almost 100 years, pure Fe_2O_3 was available and was mined in iron ranges such as in Minnesota. This hematite, 65% iron, was very soft, could be removed by steam shovel, and was shipped "as is" to the steel-producing centers. Today, hematite is scarce and the primary available ore is shifting to taconite. Taconite is a lower grade ore, usually consisting of about 25% iron, which is present as either Fe_2O_3 or Fe_3O_4.

The taconite must be concentrated; usually this is done at the mining site before it is shipped. The taconite is ground so that most of the silicates (sand) are no longer attached to the iron oxide particles. The pulverized solids are mixed with water to form a slurry; this slurry of the fine particles is then passed near a magnet. The magnetic iron particles are removed from solution while the pure sand and some nonmagnetic taconite remains in the slurry. The ore particles are thus concentrated to about two-thirds iron by weight. The iron-containing particles are agglomerated by some low-pressure pelletizing technique into about 1-in moist balls, sintered in a furnace, and then shipped to the steelmakers in places such as Gary, Pittsburgh, and Youngstown (Figure 6-2). The sintering operation consists of placing the moist pellets on a conveyor belt and then blowing very hot air through them. The pellets reach high temperature, dehydrate, and coalesce to fairly hard balls. From the grate, the pellets are taken to a kiln, where they are strongly heated and hardened.

Environmental Problems

The waste sand, or tailings (Figure 6-3), and large amounts of associated slurry water must be disposed of. Much of the waste water from the Minnesota iron ranges, for example, had until 1980 been dumped directly into Lake Superior, until court action forbid that practice. The tailing settled very slowly in the lake, forming sediments and affecting life on the lake bottom. In addition, it was found that some of the tailings released asbestos. These asbestos particles are of a small enough size (approximately that of bacteria or colloidal clay) such that they disperse completely and settle minimally.

The tailings from most iron ore concentration facilities in the United States are disposed of by landfill. This in itself creates problems. A pile of particulate waste

FIGURE 6-2. Ore carriers to transport the concentrated ore pellets from the mining area in northern Minnesota to the steel-making centers south and east.

FIGURE 6-3. Spent tailings at Reserve Mining that were, until 1980, released into Lake Superior.

118

can slide, with disastrous results if located in populated areas. Heavy rains have been known to cause this type of loss of stored solid wastes. Winds can also pollute the air with the fines from tailings piles. Some can leach, forming H_2SO_4, though that is not as serious a problem as at the mine site. And, of course, these enormous amounts of wastes are soon going to fill many of the currently available sites.

Studies are now being made as to the utilization of these wastes, but little is expected in the near future. Instead, efforts are being concentrated on methods to prevent the materials from becoming pollution sources.[6] One method that appears successful is the combined program of chemical-vegetative stabilization. The tailings themselves contain no nutrients, have poor soil texture, and contain many fines. A resinous additive is needed to first stabilize the movement until winter wheat, wheat grasses, legumes, and similar growth can become established. Fertilization is also required.[7]

Reserve Mining[8]

The legal battle involving Reserve Mining Company (Now, Cyprus Northshore Mining Co.) of Silver Bay, Minnesota illustrates not only the legal application of some of the current pollution regulations, but also the cost and the changes many industries must make in order to comply.

The Reserve Mining Company, owned jointly by Republic Steel Corporation and Armco Steel Corporation, had processed taconite (from Babbitt, Minnesota) into iron ore pellets at Silver Bay, on the North Shore of Lake Superior, since the fall of 1955. The concentration plant (Figures 6-4 to 6-6) had capacity of 10.8 million tons of pellets annually.

FIGURE 6-4.   The Reserve Mining ore concentration facility.

FIGURE 6-5.   Settling basins located at Reserve to aid in the pollution control.

The process employed at Reserve is typical of most others, and involves: 1) crushing of taconite to less than ¾-in pebbles; 2) wet grinding by rod mills to a muddy sand, the particles being of ¹⁄₁₆ in or less diameter; 3) magnetic separation; 4) classification during which pieces less than 0.004 in are passed and the larger ones are diverted to, and further ground in, a ball mill; 5) a second magnetic separation; 6) vacuum filtration of the water from the magnetite; 7) balling of the thickened magnetite into small, round pellets; and 8) baking of the pellets to a hard finish, simultaneously converting the magnetite to hematite.

FIGURE 6-6.   Reserve Mining Company operations.

The Reserve controversy arose because of the disposal of the tailings, two tons of which are produced per ton of iron ore pellets. These tailing vary in size from 1/2-in crushed rock to extremely fine sand. In 1947, the Minnesota Water Pollution Control Commission and the U.S. Army Corps of Engineers granted permits to Reserve to deposit the tailings from the taconite processing plant at Silver Bay into Lake Superior. These tailings were transported in the water. About one-half, the coarser fraction, promptly settled, and has formed a delta in front of the plant. The finer tailings, on the other hand, remained suspended, entered the bulk waters of Lake Superior and formed a heavy density current. This heavy density current settled downward and deposited the fine tailings over the bottom as it flowed toward the Great Trough of Lake Superior.

The legal controversy began in 1969. Reserve initially filed an action in Minnesota District Court, contending that the State water regulation, WPC-15, which would prevent further discharge of tailings, was unreasonable. The State filed a counter-suit, alleging pollution.

The trial, before Minnesota District Court Judge C. Luther Eckman, was completed by December 15, 1970. Judge Eckman concluded that "after 15 years of operations and discharge of tailings into Lake Superior by the Appellant, the evidence before the Court establishes that said discharge has had no measurable adverse or deleterious effects upon the water quality or use of Lake Superior insofar as its drinking water quality, any conditions affecting public health, affecting fish life or the reproduction thereof, or any interference with navigation."[9] But he also required that "the present method of discharge of tailings from its plant at Silver Bay, Minnesota shall be altered and modified by Appellant, Reserve Mining Company, to the extent that the disposition of fine tailings into Lake Superior and the distribution thereof into areas outside of, to so-called 'great trough' is discontinued."[10]

The Minnesota Supreme Court upheld this decision on August 18, 1972.

Reserve then developed and presented its plan to modify the tailings discharge, in accordance with Judge Eckman's order. The plan was to pump the tailings as a thick slurry through submerged pipes, at a depth where the fine tailings would not be affected by surface currents, wave action, or layering of water temperatures. This plan was not accepted by the State of Minnesota.

After further negotiations and other attempts to find a suitable disposal method, the federal government filed suit against Reserve on February 18, 1972. At this time, the government claimed for the first time that Reserve's tailings were similar to amosite asbestos.

Asbestos is a complex combination of several hydrous silicate minerals that contain primarily calcium, magnesium, sodium, iron, and aluminium. It is postulated that its presence in drinking water can cause tumors in the abdomen lining or the lung lining, with a latency period of about 20 years. It is likely that the size, shape, and chemical structure of the particle are important contributors to its adverse behavior. Though no evidence was available to show the tailings discharge or plant air emissions were ever a health hazard, Judge Miles W. Lord ordered, on April 19, 1974, that the Corps of Engineers provide to all affected communities "fiber-free" drinking water. Then, late on April 20, 1974, Judge Lord issued an injunction ordering Reserve's discharge to cease immediately, by midnight, due to the potential health problems. Reserve had to shut down, although two days later it was granted a temporary stay so that it could reopen the plant and mining operations.

This stay was later extended, provided that Reserve prepare and implement an acceptable plan to abate its air and water discharges. This initiated much investigation into possible landfill sites. After many more legal arguments and scientific studies, it was finally determined that Reserve dispose of its tailings by landfill at a site ("Mile Post 7") located some 7 rail miles west of the Silver Bay plant. The construction of this site was begun prior to June 1, 1977, and the tailings discharge into Lake Superior ceased by April 15, 1980, as ordered by Eighth District Court Judge Edward Devitt.

On May 27, 1977, the Minnesota Supreme Court stated, in conjunction with its final approval of the Mile Post 7 Site, that:

> It is difficult to conceive of more stringent conditions for guaranteeing acceptable air and water quality than those which have been imposed on Reserve, Amoco, and Republic and to which they have formally agreed. The permit to be granted under Minn. St. 116.081 must be reviewed and renewed every 5 years. The three companies agree to assume all risks and liabilities arising out of the operation of Mile Post 7. They are committed to perpetual maintenance of the site to prevent tailings from reentering the air and water and have agreed to all of the mitigation measures we have discussed, using the best technology available. In addition, the companies agree to comply not only with existing laws and agency regulations but those which may be adopted in the future.[11]

As part of the project, Reserve also agreed to install fabric bag filters in the railroad car dumper, the crusher building, and the storage bins. Twenty-eight wet wall electrostatic precipitators likewise were installed in the pelletizing department to eliminate any potential air pollution problems.

In addition to the tremendous legal expenses that were encountered over a nine-year period, Reserve invested more than $330,000 daily on developing Mile Post 7. The total estimated cost of the environmental control program was $370,000,000.

The Mile Post 7 disposal site has an ultimate area of 5.8 square miles. At a production rate of 9 million tons of pellets per year, the site had a design lifetime of 40 years. The plan called for coarse tailings to be hauled to the site by rail, and the fine tailings and water carried by a pipeline.

The entire site was originally a swamp and valley. In order to make the site suitable, it was necessary to construct five dams, with a total length of 26,700 ft. Initially, to minimize water inflows, a 1800-acre basin was developed. Any liquids that seep outward from the collection basin are pumped back, thus retaining all potential pollutants. The landfill began to retain water in Fall 1979, to be used as freeboard to completely cover the tailings, as was planned. Tailings themselves were first disposed of at the site in Summer 1980. Since then, the landfill has met all performance expectations.

Due to both general market conditions and partner debt, Reserve went bankrupt in 1986. The assets were acquired by Cyprus Mineral Co. in June 1989. The newly Cyprus Northshore Mining Co. began to process ore in January 1990. Under Cyprus ownership, the pellet capacity is 4 million tons annually. Even if this capacity is achieved, this disposal site should be useful for greater than 80 years.

BLAST FURNACE TREATMENT

The concentrated ore, after being shipped to the steel-making plants (Figure 6-7), is taken to blast furnaces where it is cooked with coke and limestone in the proportion 1 ¾ ton iron ore: ¾ ton coke: ¼ ton limestone: 4 tons air.

Coke

Coke is produced on-site by heating coal for 13–14 hours in tall, narrow "coking ovens" in the absence of air. Many grades of coal are mixed to obtain coke with the correct chemical and physical characteristics. The gases are driven out of the

FIGURE 6-7.   Exterior of steel mill. (*Photo courtesy DOE.*)

coal, and are recovered as coal tar, ammonium sulfate, light oil, and coke-oven gas, which eventually get converted into substances such as explosives, fertilizers, synthetic rubber, and medicines, or are used for fuel. The solid remaining is coke. This coke is quenched as it leaves the oven by the use of water, often wastewater from another step in the process. The purpose of the coke in iron and steel production is twofold: 1) to convert the Fe_2O_3 and Fe_3O_4 to pure iron (Fe) and 2) to remove impurities. In the blast furnace, the coke burns with air at about 1500°C. In limited O_2, the coke can react: Coke + $\frac{1}{2}O_2 \rightarrow CO$. The CO is a reducing agent that can react with the Fe_2O_3 and Fe_3O_4 to produce Fe. In addition, the heat melts the iron and the impurities; the latter float and are skimmed off as the white hot liquid iron flows out the furnace.

Limestone

The purpose of the limestone is to help remove the impurities in the ore. The limestone (primarily $CaCO_3$) can act as a flux, causing the impurities—particularly the silicon dioxide (SiO_2)—to melt at lower temperatures than they otherwise would. When $CaCO_3$ is heated strongly, it decomposes: $CaCO_3 \rightarrow CaO + CO_2\uparrow$. This process is called calcination. The CO_2 thus generated can react as $CO_2 + C \rightarrow 2CO$, to form CO, which acts as a reducing agent. The CaO, on the other hand, can react with SiO_2 to form $CaSiO_2$. This $CaSiO_2$ is the principal component of the solid waste slag. This slag can be separated from the molten iron because it is lighter and floats on the iron surface.

Blast Furnace Operation

In a blast furnace, the coke is fed into the top, along with the iron ore and calcium carbonate, and the mixture is "blasted" from below by hot air. The coke located near the bottom has adequate oxygen to react completely to form CO_2. As the CO_2 produced rises in the furnace, it encounters more coke and a greatly reduced oxygen supply. The CO_2 then reacts further with the coke, forming CO: $CO_2 + C \rightarrow 2CO$. Of course, additional CO is produced by reaction of the coke with the limited amount of O_2 that is available: $C + \frac{1}{2}O_2 \rightarrow CO$. The CO can react with the iron ore, to purify it, as $Fe_2O_3 + 3CO \rightarrow 2Fe + 3CO_2$. The Fe thus produced is called crude or pig iron.

Figure 6-8 illustrates a typical blast furnace. Basically, a blast furnace is a large vertical steel cylinder, lined with heat resistant "fire brick," and encircled by plates for the flow of cooling water. Common dimensions would be 150 ft tall and 30 ft in diameter at the base.

The raw materials are stored in bins near the furnace. To fill the furnace, these materials are carried to the top by 5–10-ton shuttle cars that travel on an inclined ramp. Steel-making is a continuous process; thus, the raw materials are added continuously.

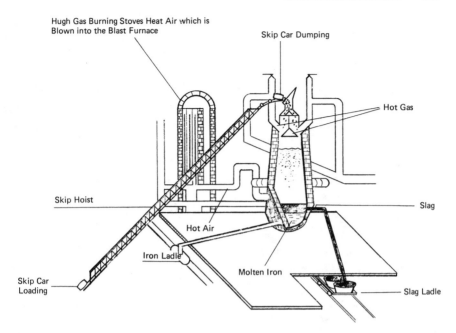

Hugh Gas Burning Stoves Heat Air which is
Blown into the Blast Furnace

Skip Car Dumping

Hot Gas

Skip Hoist

Slag

Hot Air

Iron Ladle

Molten Iron

Skip Car
Loading

Slag Ladle

FIGURE 6-8.   Schematic of a typical blast furnace.

The hot air is "blasted" into the lower part of the furnace at a rate of more than 100,000 ft³/minute. Each blast furnace is equipped with several stoves that provide this hot air. The stoves are heated to high temperatures, and then air is pumped through, simultaneously heated to 2800–3000°F and blown into the blast furnace. The stoves alternate: while one is being emptied of air, the others are heating. This air supplies both the oxygen and heat necessary to initiate and sustain combustion.

The hot air is blasted into the furnace through holes (called tuyeres) located near the bottom; the hot air then ignites the coke, causing the raw materials to melt and flow together. As the materials coalesce, the volume decreases, allowing for more raw materials to be fed in at the top. The pure melted iron, being denser than the impurities, settles to the very bottom of the furnace (to the hearth), and the slag floats on top of this 4- or 5-ft deep pool of iron. Both of these liquids remain below the tuyeres. Figure 6-9 illustrates the temperature profile of a blast furnace and the reactions that occur as the materials progress downward.

Every four or five hours it is necessary to tap the furnace to draw off the molten iron. This is done by burning out a very heat resistant plug located near the floor, at a point called the "iron notch." The iron flows in a cement trough to ladle cars, which can each hold up to 160 tons of molten iron. After tapping, and the typical removal of about 400 tons, the furnace is replugged.

It is also necessary to frequently draw off the waste slag. This procedure is

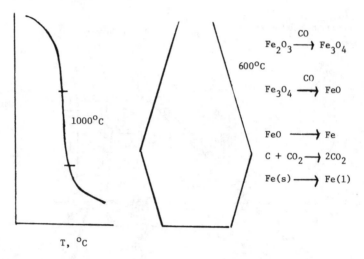

FIGURE 6-9.   The temperature profile of a blast furnace.

similar to that for the molten iron, the plugs being located higher on the furnace, above the iron layer, at the "cinder notch."

As mentioned, the blast furnace operation is continuous. Lighting the furnace is done usually by employing wood to initially ignite the coke, and with a less than normal amount of raw materials. Once lit, the furnace is shut down normally only when brick problems arise: the high heats eventually affect the brick and create hot spots. If these areas were not rebricked, serious problems would soon arise that would affect the steel exterior of the furnace. Temporarily, these hot spots can be controlled by external cooling water, but this solution is not permanent. It is hoped that a furnace can operate six or seven years without shutting down for a couple of months for rebricking.

Pollutants Associated with Blast Furnace Operation

The major pollutants associated with this stage in the iron-making process fall into four categories: air and water pollutants from the coke manufacture, and air pollutants and solid waste from the blast furnace operations. Coke manufacture also produces some solid waste. The coke ovens primarily remove organics from the naturally occurring coal by distilling off the volatile substances in the absence of air. The majority of the organic substances are collected and condensed, and recovered as a source of aromatic compounds and various other hydrocarbons and NH_3.[12] Unfortunately, not all of the emissions are ever collected. SO_2 particulates, and some organics are emitted into the atmosphere, particularly during charging (filling the oven), from door and lid leaks, and from quenching the coke with wastewater. The coke oven gas is often burned for fuel elsewhere in the plant, and can also produce these emissions. The emissions can be controlled by removing the H_2S from the coke oven gas, oven lid, and door maintenance; putting baffles in the quenching towers (and using clean, not waste, water for the quenching process); stack gas cleaning of the SO_2; using negative oven pressures during charging; or possibly hooding many of the steps.[13] The coke ovens are currently one of the major environmental concerns of the steel industry, since these emissions are very difficult to control. The water pollutants, on the other hand, are generated by coal recovery, especially by decanting the light oils. These wastes average 44 gallons/ton of coal feed, and contain phenols, NH_3, cyanides, chlorides, and some sulfur compounds. These wastes can be treated biologically either on site or with municipal sewage, and then chemically oxidized, followed by carbon adsorption for maximum pollution control.[14] In addition, some solid, very hard and fine "coke breeze" is also formed. Its typical chemical composition is illustrated in Table 6-1. This coke breeze is difficult to recycle. Some companies have succeeded in mixing it with the other waste materials such as flue dusts and $CaCO_3$ or dolomite, and then sintering it. This sintered mixture can then be used as blast furnace feed. This material can be processed very efficiently, because it contains its own fuel (coke breeze), it is already partially reduced (flue dusts), and the fluxing agents are already calcined, minimizing the energy requirements.

The blast furnace operation produces both air pollutants and solid waste. The air pollutants are primarily SO_2 and CO. The sulfur remaining in the coke oxidizes to form SO_2, which is present in the gaseous effluent conducted from the top of the furnace. The SO_2 can be minimized by the initial use of low-sulfur coal. Unreacted CO also is present in the effluent. Usually it is oxidized to CO_2 by

TABLE 6-1   Chemical Composition of Coke Breeze[a]

Substance	Percent
Total Fe	0.71
FeO	—
SiO_2	3.58
Al_2O_3	2.39
CaO	0.44
MgO	0.1
S	0.8
P	0.03
C	—
Zn	—

[a]Coke breeze analysis: 3.48 V.M., 84.99% fixed C, 11.34% ash, 0.45% moisture.

O_2 as it is emitted from the system. Part of the emitted gas could be piped to the air heating stoves or soaking pits, and the CO used as a low-BTU fuel. Sometimes potash and iron oxide can be recovered from the gaseous effluent. The slag (the solid waste consisting of the ore impurities, especially the $CaSiO_2$ that is formed) has only limited commercial utility. Some, however, can be used as a raw material for the production of cement and insulating materials. Though the percentage of the ore that forms dust is not too large (typically 1%), the large quantity of crude iron produced annually means that large piles of these wastes are common near many steel plants. Table 6-2 lists the chemical composition of typical blast furnace dusts.

Direct Reduction[15]

For many years, studies have been made as to the possibility of bypassing the blast furnace step totally, and reducing high iron concentration ores directly. A direct reduction process generally involves drying, crushing, and preheating the ore,

TABLE 6-2   Chemical Composition Blast Furnace Flue Dust and Filter Cake

B.F. Flue Dust	
Substance	Percent
Total Fe	33.8
FeO	16.6
SiO_2	7.5
Al_2O_3	1.57
CaO	3.95
MgO	3.83
S	0.34
P	0.09
C	34.1
Zn	0.15
B.F. Filter Cake	
Total Fe	32.5
FeO	10.7
SiO_2	4.88
Al_2O_3	2.44
CaO	6.81
MgO	2.87
S	0.39
P	0.09
C	28.5
Zn	0.97

which must contain about 68% Fe, and then injecting the ore into a reducing column. One of the most efficient reducing gases is H_2. This converts the iron oxides directly to metallic iron, which can then be further processed in electric furnaces.

There are several definite advantages for direct reduction of Fe. The blast furnace consumes approximately 50% of all the energy used in the steel production. The blast furnace process also adds carbon to the Fe, and this must subsequently be removed. In addition, the blast furnace generates some of the most serious environmental problems of the industry.

Interest in direct reduction methods revived between 1965 and 1970. Earlier schemes had proven not to be economical. The trend toward smaller facilities, often producing specialty steels, enhances the applicability of the general process.

A number of companies in other countries are developing a wide variety of direct reduction techniques to produce iron for steelmaking. These processes include the following:

- The original direct reduction process, the Hyl process, was developed several decades ago by Hylsa S.A. (Monterrey, Mexico). Natural gas is reformed to hydrogen and carbon monoxide, reducing gases that, at 800-900°C, are used to strip oxygen from the ore to obtain sponge iron.
- The Corex process, a German process used commercially since 1989 at Iscor Ltd's Pretoria Works in South Africa, reduces the iron ore in a shaft furnace. The reduced ore is then melted in a "melter gasifier," which produces the reducing gases for the shaft furnace.
- The Japanese Iron & Steel Federation is developing a pressurized process in which ore pellets are fed into a fluidized bed reduction furnace into which coal is injected from the bottom and oxygen is injected through a lance. The partially reduced ore and tar are then fed to a smelting reduction furnace.
- The Fastmet process uses finely crushed iron ore, coal, water, and a taconite binder, which are pelletized in a drum. The pellets are partially pre-dried with air. Overfire burners heat the pellets sufficiently to gasify the coal to produce reducing gases, rather than firing the pellets directly at high temperature.
- The Hismelt process, a joint Australian-U. S. project, uses injected coal and oxygen to produce iron. Any surplus CO from the reactor is used to preduce ore fines in a fluidized bed.

The key environmental advantage to all of these processes is to avoid the use of coke. A number of steel companies, rather than installing direct reduction processes at this time, are thus modifying their blast furnaces to reduce coke use. This can be done by injecting coal or other fuels into the combustion zone. Coal injection can, for example, reduce the coke used from 900-1000 lb/ton of hot metal to 500-600 lb/ton.

It remains to be seen whether this improved technology will reverse the century-old economic advantage of indirect processing.

STEEL PRODUCTION

The product of the blast furnace operations is "pig iron", a combination of about 95% iron, 3-4% carbon, and small amounts of manganese, phosphorus, sulfur, and silicon. The name originally arose because the iron is initially cast into bars called "pigs." This pig iron was once cast in molds shaped like little piglets surrounding a main channel, the sow. The majority of the pig iron is converted to some form of steel. Smaller percentages are used to produce cast iron and wrought iron, processes usually conducted as part of foundry operations.

The purpose of the steel production stage is twofold: 1) to remove impurities, primarily the excess carbon, and 2) to add desirable materials to form alloys. Pig iron is not a good structural material due to its low strength, caused primarily by the presence of carbon impurities. These steps thus are aimed to enhance the strength and durability of the product. The specific properties of various types of steel can also be influenced by the type of heat treatment steps to which it is subjected. For example, after the steel undergoes rapid cooling, or quenching, it can be reheated to a lower temperature to "temper" it (to increase

FIGURE 6-10.   Steel mill operations. (*Photo courtesy DOE.*)

its strength and ductility). If the steel has been hardened too much in rolling, it can be resoftened by annealing—heating the steel, followed by a very slow cooling.

There are several general types of steel that can be produced:

1. *Carbon steel.* This is the most common type of steel, produced primarily by removing the carbon from the pig iron until less than 1% remains. The exact amount of carbon remaining largely determines its properties. Much of the carbon steel is used as structural steel, and as such is about 0.25% carbon.
2. *Alloy steel.* Though this type of steel does also contain small amounts of carbon, the properties are determined for the most part by the other additives. In this category, we find the "stainless steels," which strongly resist corrosion due to the presence of 11-27% chromium. This chromium forms a protective, self-renewing film of chromium oxide (Cr_2O_3) on the surface. This Cr_2O_3 layer prevents the formation of rust ($Fe_2O_3 \cdot xH_2O$) on the steel surface. The chromium can also increase the hardness of the steel and makes it quite resistant to both acids and alkalies. In terms of carbon content, stainless steels range from 0.08-1.2%. Many may also contain some nickel. Other typical alloying materials and the properties they can impart to steel are:
   a. *Nickel*—increases toughness; increases resistance to acids and to high temperatures.
   b. *Tungsten*—increases hardness at high temperatures.
   c. *Molybdenum*—increases strength; improves ability to resist heat.
   d. *Manganese*—increases strength; improves wearability.
   e. *Vanadium*—increases strength and resiliency.
3. *Tool steel.* As the name indicates, this is the fine grade steel used for tool production. The properties can vary significantly. Some tool steels use only carbon, others use one or more of the alloys to control the desired characteristics.

Specialty steels—that is, stainless and tool steels, nickel- and cobalt-based high-temperature alloys, magnetic alloys, electrical steels, and related products—are of importance to the U.S. economy. They are relied upon for applications in aerospace, manufacturing, power generation, chemical and oil processing, defense, electronics, food processing, and many other aspects of daily life. They are also a major factor for the U.S. steel industry, itself, for though they represent only 1.5% of the total tons of steel consimed in the United States annually, they represent 10% of the total sales value.[16]

Steel is manufactured using not only pig iron as the raw material, but also scrap iron and scrap steel. These "waste" materials currently supply about 50% of the total raw material needs for U.S. steel production.

The steel-making process itself is very complex. The metal must undergo several treatment steps:

1. Melting and processing to remove carbon and/or add alloy materials (the actual "steel-making" step);
2. Rolling mill operations;
3. Pickling;
4. Final shaping; and
5. Finishing.

These operations are quite diverse, as are the environmental considerations. To clarify the source and nature of the various pollutants, the discussion of each of the above steps is followed immediately by a consideration of the environmental problems and the currently feasible solution alternatives.

Steel-making

There are four common methods of steel-making: 1) open hearth furnaces, 2) the basic oxygen process, 3) electric furnaces, and 4) the Bessemer converter.

Open Hearth Furnaces

Use of an open hearth furnace (Figure 6-11) is the classical steel production method. The furnace consists of an open, saucer-shaped floor, which is exposed directly to flames. The entire furnace is lined with refractory brick. Air and fuel (gas, oil, tar, etc.) are heated below the hearth and then blown in directly above the steel and ignited. Temperatures are at about 3000°F. After 8-10 hours, the batch of melted steel is tapped from the side, opposite from where the raw materials are added. Each batch typically weighs 50-500 tons, depending upon the furnace capacity.

The typical feed is about 5-6% $CaCO_3$, and equal portions of pig and scrap iron. Initially, the furnace is filled with the $CaCO_3$ and scrap. The $CaCO_3$ acts as a flux, as in the blast furnace operations, thus removing impurities and forming a slag. The impurities are removed by oxidation caused by the O_2 in the air and the Fe_2O_3 from the rusty scrap. The Fe_2O_3, for example, can react with the excess C as: $Fe_2O_3 + 3C \rightarrow 2Fe + 3CO \uparrow$. The limestone will react with the sulfur and phosphorus oxides, and prevent their emission:

$$CaO + SO_2 \rightarrow CaSO_3$$
$$6\,CaO + P_4O_{10} \rightarrow 2Ca_3(PO_4)_2.$$

In addition, it will react with the silicon and manganese oxides, the products of which form part of the slag. After about an hour, the molten pig iron is added. By this stage, much of the remaining carbon in the iron has boiled off as gaseous CO_2. After the desired amount of carbon is removed from the mixture, manganese-containing "Spiegeleisen" can be added to help remove some of the entrained O_2.

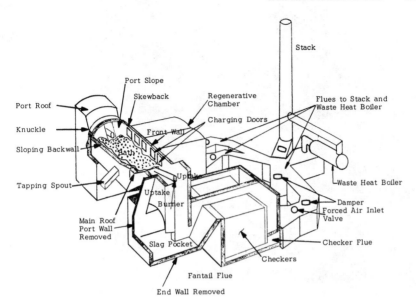

FIGURE 6-11.   Schematic of an open hearth furnace.

Spiegeleisen is typically 16–28% manganese, less than 6.5% carbon, and 1–4.5% silicon. The use of Spiegeleisen has decreased in recent years. After about a half hour, as the furnace is tapped, ferromanganese is added to the steel to further remove more O_2. Ferromanganese is available in several grades, but typically contains 74–82% manganese, not more than 1.25% silicon, 7.5% carbon, and small amounts of sulfur and phosphorus. The tapping procedure allows the molten product to overflow the ladle so that the slag runs over into a separate ladle called a slag thimble; the steel is then poured into ingot molds (Figures 6-12 and 6-13). At this stage, small amounts of aluminum may be added to further deoxidize the steel. The slag must, of course, be disposed of. The other pollutants are primarily CO and a rusty red smoke due to particulate Fe_2O_3.

The Basic Oxygen Furnaces
The basic oxygen process (BOP) is principally the same as the open hearth process, except the system uses O_2, not air. This adaptation permits a faster and more economical processing of the materials. The open hearth process typically takes 8 hours to process steel; the BOP, 40–45 minutes. Most open hearth operations in the United States have been phased out and replaced by BOP operations.

The Electric Furnaces
The electric furnace (Figures 6-14 and 6-15) is a round steel shell, lined with fire brick. Three carbon electrodes, each approximately 6 ft long and 1 ft in diameter,

FIGURE 6-12.   Pouring steel. (*Photo courtesy DOE.*)

FIGURE 6-13.   Steel being poured into ingot molds before preliminary rolling. (*Photo courtesy DOE.*)

FIGURE 6-14. The submerged arc furnace at Roane Electric Furnace, Rockwood, Tennessee. Seven megawatts of power are supplied to three 36-in. carbon electrodes for the production of a ferromanganese alloy. (*Photo courtesy DOE.*)

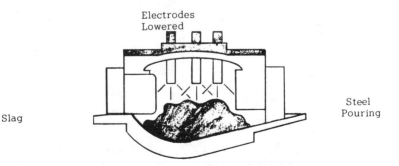

FIGURE 6-15. Schematic of an electric furnace. Electric current from electrodes provides heat to melt the scrap.

are inserted through the top. Large amounts of electricity are fed to the electrodes, the current arcs among the electrodes and the molten slag generating extremely high temperatures (3500°F).

The melting and refining operations are very similar to those of the open hearth process. The process is, however, restricted primarily to the recovery of scrap steel and alloys, not the processing of pig iron. Much of the product is a precise type of alloy or of carbon steel, so the process is used when it is essential to have a steel with very exacting characteristics.

Bessemer Converters

The Bessemer converter (Figure 6-16) is an egg-shaped, open-topped, tippable furnace. The converter is tipped on its side, filled with molten iron, and then righted. Air is then blown into the bottom, through the molten steel, at a pressure greater than 20 psi and at a rate of over 25,000 ft³/minute. This air oxidizes the impurities, and burns them out of the iron. As the air flow is initiated, flames and sparks fly upwards of 30 ft, changing from red to yellow to white as the manganese, silicon, and then carbon are oxidized. The flame dies back after about 10 minutes, at which time the steel is poured from the converter. A modern modification of the Bessemer process is to use O_2, not air, in order to eliminate problems that are created by the N_2 present in the air. This process reduces the processing time for the steel to 20–25 minutes.

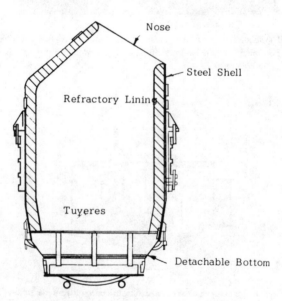

FIGURE 6-16.   Schematic of a Bessemer converter.

Steel made by the Bessemer process contains larger than average amounts of sulfur and phosphorus. The result of this is a steel that is stiffer, harder, and more easily machined or threaded than similar BOP steel.

Pollution Problems Associated with Furnace Operations

The air pollution problems of these various processes are concentrated in the fume generation of the furnaces and in the pouring operations. The furnace emissions (Figure 6-17) are the most serious environmental problem that the steel companies have to control. Open hearth and BOP furnaces usually employ electrostatic precipitators or venturi scrubbers for particulate removal; many of the newer electric furnaces employ baghouses. A carbonaceous material called kesh is also generated and emitted during the molten metal transfer operations. The appropriate areas should be hooded, and the vented air cleaned by baghouses, to control these emissions.

The water pollution problems of these stages in the steel processes result from employing wet methods for air pollutant removal. The usual problem is thus suspended solids. Usually, recirculating such waste water by employing cooling towers

FIGURE 6-17.   Air pollution generated at the Roane Electric Furnace Company in Rockwood, Tennessee prior to installation of control equipment. (*Photo courtesy DOE.*)

is the most generally satisfactory solution. The processes involving oxygen are particularly troublesome. Treating such waste waters requires chemical coagulants, and even that treatment is seldom totally efficient. A process involving magnetic addition agglomeration has proven itself fairly effective in treating BOP wastes.

The disposal of the recovered solid wastes, particularly those from the BOP, open hearth, and electric furnace operations, is very difficult. The relatively high concentrations of zinc (which may result from the scrap metal feed) are often assumed to make the material unavailable for reuse. Zinc, and also lead, vaporize near the bottom of a furnace, but as they travel upward they cool and "freeze" on the walls. This leads to deterioration of the refractory brick and eventual choking off of the furnace. There are two types of recovery methods generally considered for zinc removal: 1) an acid leach, typically by H_2SO_4 or HCl, or 2) pelletization of the dusts followed by anaerobic roasting. None of the specific methods so far developed are feasible solutions. The economics are often bad, and, in addition, there are often scientific difficulties. For example, there are problems that do arise during leaching due to the zinc ferrites that are often present; these salts are very difficult to separate from the wastes. Some plants can sinter these waste dusts with coke breeze and $CaCO_3$, or dolomite, as mentioned, but this procedure is also not satisfactory across the industry. Table 6-3 lists the typical chemical composition of various of these steel-making solid wastes.

Direct Steel-making

According to an intensive study conducted by American Iron and Steel Institute (AISI) researchers[17] the next generation of steel-making processes will be based on "inbath" smelting technology. Inbath technology is radically different from

TABLE 6-3.   Chemical Composition of Steel-Making Dusts

	B.O.F. Sludge (%)	Open Hearth Dust (%)	B.O.F. Sludge (%)	Electric Furnace Dust (%)
Total Fe	60.1	45.5	54.4	37.3
FeO	49.3	0.5	12.6	0.9
SiO_2	1.5	0.42	2.14	1.58
Al_2O_3	0.1	0.47	0.56	0.08
CaO	4.49	1.38	5.8	11.15
MgO	0.12	0.86	1.18	2.75
S	0.27	1.95	0.18	0.57
P	—	—	—	—
C	—	—	—	—
Zn	1.30	17.26	2.8	17.47

conventional treatment methods. The driving force behind its development has been based on both effluent/emission considerations and energy conservation.

The inbath smelting technology is intended to replace traditional coke ovens, blast furnaces, and basic oxygen furnaces with a direct reduction process. In contrast to other direct reduction technologies, this process would eliminate not only the majority of conventional iron-making processes, but also major aspects of steel-making.

Inbath smelting uses coal instead of coke. Eliminating the coke ovens, which are increasingly expensive to replace and operate because of environmental problems, reduces the cost per ton of steel while maintaining existing capacity. With this technology, coal and iron ore are fed into a liquid bath of iron containing a high percentage of carbon. The carbon reduces the iron by removing oxygen from the ore, forming carbon monoxide and molten, elemental iron.

Oxygen is injected into the exhaust gases before they leave the smelting vessel. Some of the gases are burned. This process, called post combustion, generates additional heat. This energy is recovered in a prereduction process. The hot, reducing exhaust gases are recycled to a prereduction vessel where they are reacted with the incoming ore before it is injected into the bath. In this way, the ore is not only heated, but also a portion of the oxygen content is removed prior to charging the ore into the smelting vessel. Alternatively, the gases could undergo further reactions to produce an even more reducing gas, or it could be recovered for a fuel gas for elsewhere in the plant.

After the inbath smelting, the molten iron potentially can be treated by a continuous refining process for desulfurization and decarbonization.

Extensive tests, sponsored jointly by the U.S. Department of Energy (77% of the costs) and AISI (23%) were initiated in 1989, regarding various aspects of the processing, to optimize heat transfer, slag utilization, and other similar considerations. One of the main limitations of the process, to date, is the inability to recycle scrap. Projections are that the direct operating costs can be reduced by $10 to $25 per ton of steel produced, based on an estimated energy cost saving of about 20%.

Rolling Mill Operations

After the furnaces, the rolling mill operations are the next step in the steel-making process. These operations consist of either conversion of steel ingots to forms called blooms, slabs, or billets, by traditional techniques, or slab production by continuous casting. ("Blooms" are rectangular blocks with a cross-sectional area of greater than 36 in²; "billets" are longer and narrower; a "slab" is wider and flatter than a billet.) (Figure 6-18.) With the conventional techniques, the partially finished steel is first, before any actual mill operations, cooled in molds after the removal from the furnace, and then subjected to a series of steps, including chipping, grinding, and scarfing (use of O_2 torches to remove surface defects). The steel ingots must

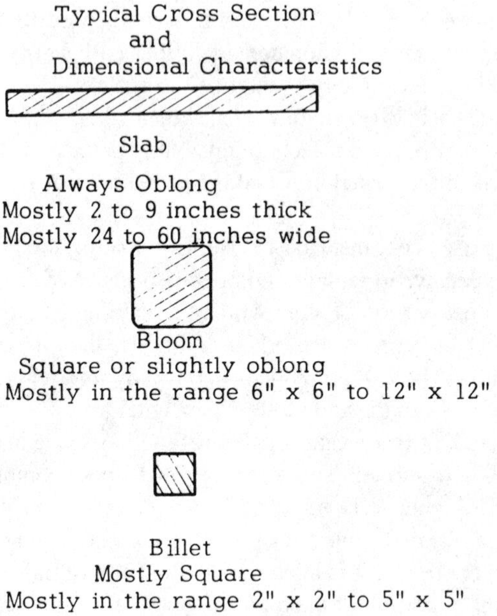

Typical Cross Section
and
Dimensional Characteristics

Slab

Always Oblong
Mostly 2 to 9 inches thick
Mostly 24 to 60 inches wide

Bloom

Square or slightly oblong
Mostly in the range 6" x 6" to 12" x 12"

Billet
Mostly Square
Mostly in the range 2" x 2" to 5" x 5"

FIGURE 6-18.   A comparison of blooms, billets, and slabs.

be heated to about 2250° F before shaping, and this is usually done in heating pits called "soaking pits." At these temperatures, the ingots become white hot and somewhat soft. The steel then goes to one of several different types of mills, depending upon the ultimate product desired. In general, there are two classes of mills: 1) semi-finishing mills, which form only rough rectangular steel blocks, and 2) finishing mills, which roll the rough steel shapes into usable products. These latter include plate mills, mills for rolling rails and structural shapes, wire mills, bar mills, tube mills, and continuous strip mills. At this stage in the steel processing, only the semi-finishing mills are relevant: the ingots are cast simply into rough blocks. Only after several more processing steps are the final products formed.

The rolling mills resemble large clothes wringers, compressing the softened ingots between heavy rolls into long, thin shapes. Mills have typically two, three, or four rolls through which the steel must pass. After an initial hot rolling, much sheet steel may be further processed by a cold (room temperature) rolling procedure.

Many plants are now instituting "continuous casting." Rather than cooling the steel in molds after the furnace operations and then reheating it prior to rolling, the molten steel is allowed to cool only slightly (to a temperature comparable to that obtained in the soaking pits) and then rolled immediately into the desired shapes. A large percentage of the steel companies use continuous casting methods now, even for specialty steels, for they significantly reduce the energy consumption.

Rolling Mill Pollution

The primary air pollutant from these operations are the airborne particles generated by hot scarfing—using the O_2 torch on hot steel. If wet collection methods are employed, this degenerates into a water pollution problem.

The chippings and so forth amount to large tonnages of solid waste, but much can be used as scrap, or can be sintered for blast furnace feed. Most of the scale and lubrication oil produced by the rolling mills is suspended in water (these operations produce about half the contamination of water from a steel plant) and can be recovered by the use of settling chambers or "scale pits." The emulsified oil generated by the cold rolling process is one of the more difficult waste water treatment problems. Much of this waste can be recycled after collection by chemical coagulation, magnetic agglomeration, or filtration. (See *Finishing Operations and Environmental Problems*, p. 144, for further discussion.)

High-grade magnetic separation is finding increasing use for removing suspended solids and regenerating steel process waters. The use of low-grade magnetic fields for separation is an old technique, but is applicable only for highly magnetic materials. High-grade magnetic separation, on the other hand, with its use of high magnetic field intensity, large magnetic gradients, and large volumes and surface areas, can be used also on materials only weakly magnetic.

In practice, a large, cylindrical drum is packed with a finely stranded (ferromagnetic) stainless steel wool, which is then placed in a high magnetic field. The contaminated water from the steel rolling operations, containing as much as 150 mg suspended solids/L, can flow through the system and have 90% or greater of the solids filtered out.

The estimated costs for such a system are about $100,000 for 1000 gallons/day with estimated operating expenses of 13.7¢/1000 gallons, both of which are lower than for a conventional treatment facility (1978 dollars). Though the first commercial installations were all in Sweden and Japan, a small unit was successfully operated in the United States at Inland Steel's East Chicago plant.[18]

Pickling

Most steel products need to undergo some further shaping after the initial rolling mill operations. Before this "final" shaping of the steel can occur, the steel must undergo a pickling process. The purpose of this step is to strip the oxide coating from the metal surface. Usually, either sulfuric acid (H_2SO_4) or hydrochloric acid (HCl) are used, though nitric acid (HNO_3) or phosphoric acid (H_3PO_3) are used for some applications, such as with stainless steel. For many years, H_2SO_4 was the major acid used, but its rising cost, combined with the decreasing cost of HCl, have made the latter the primary choice for new installations. An additional advantage is that HCl can pickle faster and at lower concentrations, and supposedly gives a "cleaner" finish. It may, however, lead to more corrosion. Which acid is used is important; disposal methods may vary greatly.

For many purposes (for example, for wire and steel strips), the pickling can be accomplished in a continuous system using several tanks in series, each succeeding tank containing acid of greater strength. The last tank in line contains fresh acid, the first contains the almost spent liquor. This latter tank is the one whose contents are disposed of. Steel bars and pipe are often pickled in batch-type operations, on the other hand. Newer systems allow lower acid usage, to minimize the environmental effects.[19]

Spent Liquor Problems

How large a problem is the spent pickle liquor? It may be the second largest problem for the steel industry, second only to the furnace particulate emissions.[20] The spent liquor typically contains 0.5–10% acid and 12% iron, both which must be removed. The acidity effects of the waste water are fairly obvious; the iron may turn the water muddy brown, form slime, and exert a strong O_2 demand.

The quantities of spent liquor are also prodigious: 8–15 gallons of spent liquor are typically generated per ton of steel produced, thus resulting in the production of one-half billion gallons/year (assuming an annual production rate of 50 million tons).

Pickling liquor disposal can be handled in several ways. It can be actually disposed of by: 1) directly discharging it to a waterway, 2) sequestering it in deep wells, or 3) neutralizing the acidity and then discharging it. Or it can, instead, undergo some type of recovery and/or regeneration process.

Direct discharge into natural water systems is perhaps the least expensive disposal technique, but also the one most likely to cause serious environmental difficulties. Several steel companies have had permission to discharge their pickling liquor into Lake Erie, due to the lake's known alkalinity, which could neutralize the remaining acidity. The iron (and its associated problems), of course, remains.

Deep well sequestering is presently one of the more popular disposal methods. If the geology of the site is appropriate so as to not allow any ground water contamination, a well may be drilled and appropriately cased, and then used for liquor disposal at a relatively low cost.

Simple neutralization of the liquor of a pH of about 7 is easily done, but does create some problems. A particular difficulty is that the solids that form (for example, ferrous hydroxide) contain large percentages of water. This sludge must be thickened in lagoons before final disposal. Since the pickling of one million tons of steel can generate 200,000 tons of wet sludge, this method is feasible only for those plants that have quite large land areas they are able to use for this purpose.[21] Sophisticated neutralization techniques are required if any recovery is desired (see below).

A more expensive method of liquor disposal is to have the waste hauled away by a contractor. The contractor would then be responsible for the ultimate disposal of the wastes, or the recovery of any useful byproducts. There are non-scientific problems that may result for the steel company from this procedure unless the contractor is reliable.

Recovery processes can be designed to either recover just the free acid or to also regenerate the acid from the iron sulfates or chlorides that have formed. Recovery of only the free acid is practical on H_sSO_4 systems, where the free acid level may be as high as 10%, but not usually on HCl systems, where it can be as low as 0.3%.

One free acid recovery system requires gradual cooling of the H_2SO_4 liquor from around 180°F to the ambient temperature. This allows the formation and precipitation of iron sulfate heptahydrate ($FeSO_4 \cdot 7H_2O$), green crystals that can easily be separated from the free acid. The $FeSO_4 \cdot 7H_2O$ can, in some cases, be marketed as a flocculent for sewage treatment plants or for the production of paint and fertilizers. Otherwise, it can be further treated by either roasting the crystals to produce Fe_2O_3 and gypsum (hydrated $CaSO_4$), which also has commercial utility.

The recovery of not only the free acid but also that incorporated in the various iron salts, "regeneration," does have definite advantages, but due to the increased installation and operating costs of these more complex systems, many steel mills have not chosen to apply them.

There are several regeneration processes that are available for both H_2SO_4 and HCl systems:

1. The "DuPont Process" involves controlled neutralization of the sulfuric acid by CaO, and air oxidation of the mixture. In contrast to simple neutralization of the acid (a process that produces sludge that is difficult to dewater), this process generates a small volume of easily dewatered sludge consisting principally of Fe_3O_4 and gypsum. The Fe_3O_4 can be magnetically separated from the mixture and used as blast furnace feed. The gypsum can be marketed for wallboard or sold to the cement industry for cement manufacture. Any other solids that are produced would be landfilled.

2. Another possible process involves the roasting of the $FeSO_4$ produced in the H_2SO_4 system with carbon and O_2. This produces FeO and SO_2, both of which can be recovered. A comparable process, for use with HCl systems, is to roast the liquor to drive off HCl gas, which can then be absorbed in water to produce the acid. Also produced in the roasting is Fe_2O_3, which is finding some markets as a paint pigment and in the manufacture of magnetic tapes.

3. A process involving evaporation and crystallization can also be used in H_2SO_4 processes. Evaporation of the liquor at 160°F, followed by crystalization of $FeSO_4$ in a vacuum crystallizer, cooling the solution to 30°F, and then centrifuging, produces a liquid that can be diluted to about 25% acid and returned to the pickling step.

4. A fairly new method for recovery in HCl systems is the direct addition of H_2SO_4 to the HCl liquor. The result is ferrous sulfate heptahydrate crystals, which can be recovered as before, and HCl for reuse as the pickling liquor.
5. Dialysis is a technique that has been used particularly for stainless steel pickle baths. It is a separation process utilizing diffusion as its driving force. Because no power or chemicals are used, the operating cost is minimal. It can be used to recover raw materials—often 70–75% of the currently wasted acid—and can significantly reduce plant wastes.[22]

How do the various steel companies actually handle their waste pickling liquor? This varies not only from company to company, but also from mill to mill. Several of the various factors which have an influence are the age of the plant, the use of either H_2SO_4 or HCl, the exact environmental regulations and conditions in the locality in which the plant is situated, and the professional opinions of company management. Some industries still prefer to dispose of their wastes by deep well injection. Others use neutralization or recovery processes. All trends indicate that in the near future, recovery and regeneration will become more and more necessary and prevalent.

Finishing Operations and Environmental Problems

The final steel finishing operations consist of processes such as final rolling in finishing mills, tin plating, galvanizing, chrome plating, coating, tempering, and/or polishing. The air pollution problems associated with these steps are negligible. The solid wastes are more significant and consist primarily of scrap and sludges or precipitates generated by various wastewater treatment methods.

The treatment of the wastewaters is the most important environmental issue of these final operations. Of particular difficulty is the emulsified oil from any cold rolling operations, as was experienced in the earlier rolling operations. The emulsification is what causes the difficulties. The various treatment methods available include the use of emulsion-breaking agents, magnetic separators, or possibly iron salts and CaO followed by air flotation. Often, plants find it preferable to leave the removal of this oil to outside contractors.

The hot strip rolling process also creates water problems. Throughout this process, the steel slabs, sheets, blooms, billets, or bars are being oxidized, cooled, and washed by a high-pressure water spray. This procedure creates (on the surface) a layer of metal oxides called mill scale (Table 6-4). During the rolling operations, this scale is continually broken away and then reformed. The scale itself falls through the roll tables into a sewer with a flowing water stream. Also into this sewer fall the lubrication greases and oils from the rolling machinery and other mill debris. These substances—the mill scale, grease, oil, water, and other debris—form a sludge. This sludge comprises approximately 12% of the total solid wastes of a steel plant.

These sludges are collected in settling pits and basins. Traditionally, these

TABLE 6-4   Chemical Composition Mill Scale[a]

Subsance	Percent
Total Fe	74.7
FeO	63.6
SiO_2	0.87
Al_2O_3	0.74
CaO	0.35
MgO	0.08
S	0.04
P	0.04
C	—
Zn	—

[a]Slag analysis: 11.6% metallic Fe, 63.6% FeO, 19.5% Fe_2O_3

sludges have been removed from the collection area by cranes (or similar machinery) equipped with clam shell buckets. These larger particles can be removed and trucked back to the iron production operation for reuse. The smaller and more difficult to handle particles initially remain in the sewer water stream, and do not collect easily in the settling pits and basins. The continual turbulence caused by dredging contributes to the difficulty experienced in settling. These fine particles must be collected by filtration units and terminal lagoons. Generally, these fine particles are disposed of by landfill, not recycled.

A several-step process first used by a steel plant in 1968 is a possible method for converting this sludge to a "raw material" with an iron concentration greater than that experienced in the normal iron ore concentration steps. In addition to concentrating the iron to about 70%, this process simultaneously removes much of the oil contaminants from the mill scale. This oil had traditionally created air pollution problems at the sintering and pelletizing plants as the sludge was recycled, somewhat limiting the use of even the large sludge particles.[23]

The various plating and other metals finishing operations also generate large amounts of wastewater. The wastes and the specific operations are discussed in detail in Chapter 9.

Minimill Technology[24]

Minimills utilize electric arc furnaces to process scrap. They now account for some 20% of U.S. production and have capacities of approximately 130 ton/hour, about half the 200–300 ton/hour typical for a standard BOP furnace.

Minimills are more flexible than many traditional steel plants, and they avoid the environmentally poor coke/ironmaking steps. However, they are quite energy intensive. Much of the product metal also contains traces of other metal contaminants, which prohibits its use for sheet steel in the automotive industry. Moreover, the electric arc process produces ionized nitrogen, which can lead to excessive hardening of the steel.

A number of new processes are now being investigated for minimills. Studies have focused on new thin slab processes (for sheet metal production), processes to avoid hot rolling, and energy-saving casting that can reduce slabs immediately and wind them into coils. Hopefully, techniques will be developed that can save energy and/or reduce contamination.

Other Environmental Considerations

Steel processing requires large amounts of coal for coke production and for fuel used elsewhere in the plant. Many steel companies own their own coal fields, and mine them to obtain the necessary coal. U.S. Steel owns, for example, approxi-

FIGURE 6-19.   Cladophora growing on shore adjacent to typical thermal plume outlet. (*Photo courtesy Wisconsin DNR.*)

mately 25% of the coal fields in the United States. The majority of the coal must be obtained from underground mines in order to be of a metallurgical grade. This mining, naturally, also has a major impact on the environment, and is discussed further in Chapter 8.

Thermal pollution, from the furnace cooling water, is another environmental consideration (Figure 6-19). Typically, the cooling water is released only 6-7°F above its intake temperature, well below the 18°F rise in temperature often experienced with power plant operations.

Ferrous products usually constitute 99% of the total metals found in trash. Some communities and many industries are recovering their ferrous wastes (Figures 6-20 and 6-21); collection by use of magnets is relatively easy. Reuse is sometimes difficult if the materials are contaminated by even trace amounts of some other metals, such as copper, lead, and zinc. Automobiles, for example, are difficult to recycle totally. Automobile engines and radiators generally are not recycled. The market value of scrap ferrous metals varies by more than a

FIGURE 6-20.   Recycled steel from vacant buildings is being used in the construction of PPG Industries, Fresno, California glass plant. (*Photo courtesy PPG Industries.*)

FIGURE 6-21.   Scrap metal that is being collected and will eventually be processed.

factor of two when comparing times of high demand to times when the demand is low.

References

1. Parkinson, Gerald, "Steelmaking Renaissance." *Chemical Engineering* **98**, *No.5*, May 1991, pp. 30–35.
2. "Digging Into Mine Waste." *Environmental Science and Technology* **8**, *No. 1*. February 1974, p. 111.
3. *Ibid.*
4. *Cleaning Our Environment—The Chemical Basis for Action.* Washington, D.C.: American Chemical Society, 1969, p. 114.
5. *Pollution Control Technology*, Research and Education Association, New York: 1973, p. 453.
6. Bramer, Henry C., "Pollution in Steel Industry." *Environmental Science and Technology*, **5**, *No. 10*, October 1971, p. 1004.
7. "Vegetation Tames Mine and Smelter Wastes." *Environmental Science and Technology* **11**, *No 5*, May 1977, p. 462.
8. "Reserve Mining: The Agreement to End Taconite Discharge to Lake Superior." *Environmental Science and Technology* **11**, *No. 10*, October 1977, p. 948.
9. "The Reserve Mining Company Controversy." Printed by Reserve Mining Company, p. 3.
10. *Ibid.*
11. *Ibid.*, p. 18.

12. McGannon, Harold E., *The Making, Shaping, and Treating of Steel, 8th Edition.* Pittsburgh: U.S. Steel Corporation, 1964, p. 121.
13. "Stage Charging Reduces Air Emissions." *Environmental Science and Technology* **8**, *No. 13*, December 1974, p. 1062.
14. Bramer, "Pollution in Steel Industry." p. 1004.
15. Parkinson.
16. Roedel, Paul R., "The U.S. Specialty Steel Industry." *ASTM Standardization News*, **15**, *No. 11*, November 1987, pp. 52–53.
17. AISI Direct Steelmaking, Pittsburgh, PA: AISI, 1989.
18. "Magnetic Separation Finding New Uses." *Chemical Engineering News*, July 10, 1978, p. 26.
19. Roedel.
20. "Water Pollutant or Reusable Resource?" *Environmental Science and Technology*, **4**, *No. 5*, May 1970, p. 380.
21. *Ibid.*
22. *Pollution Control Technology,* p. 450.
23. "Steel Industry Sludge is Being Reused." *Environmental Science and Technology* **9**, *No. 7*, July 1975, p. 624.
24. Parkinson.

7

Foundry Operations

The foundry (Figure 7-1) is the next stage of processing for the majority of that portion of the pig iron produced in blast furnaces that is not converted to steel. In addition, some foundries cast steel, and some brass, bronze, magnesium, and/or aluminum. Foundries differ greatly in the size of operation, the size of typical castings, the type of materials cast, the types of molds and cores, and even the

FIGURE 7-1. Typical foundry emissions due to melting by cupola if not controlled.

degree of automation, but in spite of these differences, they all have the same major departments. These include the following:

1. Melting
2. Pouring
3. Molding
4. Cast-cleaning
5. Core-making
6. Pattern-making
7. Casting repair
8. Maintenance.

PROCESSES AND POLLUTION

The major environmental problem associated with foundries is that of air pollution. Every major foundry in the United States, without exception, has installed some type of air pollution control equipment. But the problems span the entire range of recorded and recognized pollutants.[1]

All modern foundries employ internal dust ventilation systems to maintain high-quality internal ambient air (Figure 7-2). This ventilating air cannot be directly exhausted to the surroundings any more than can the exhaust gases from the melting processes.

FIGURE 7-2.   Ductwork exhausting individual work stations in the chipping and grinding area.

The composition and size of the dusts from the various operations do vary significantly. The dust from the grinding of castings is primarily coarse particles, consisting of (for ferrous operations) elemental iron, iron oxide, and clay. On the other hand, the dust from a castings "shakeout" is very fine carbonaceous material. This dust is normally entrained in a moist air stream.

There are, in addition, gaseous pollutants that form. Oils, resins, and other

$$\overset{\displaystyle H}{\underset{\displaystyle |}{}}$$

organic materials used in the processing form aldehydes $(R-C=O)$, unsaturated hydrocarbons, and other odorous compounds when distilled or pyrolyzed. Sulfur is an additional objectionable vapor. The majority of these types of compounds are emitted from the core-baking ovens. Other sources are from the oil-vaporizing furnaces used for metal chips prior to the metal-melting operations, the paint-baking ovens, shell molding, and similar procedures.

Because of the wide variety in pollutants whose characteristics depend upon the particular operation and process, it is convenient to consider each step in the process and its respective pollutants separately.

Melting Department

Ferrous Operations

The metal is first sent to the melting department, where it undergoes some form of heating and melting. There are several common heating methods used, including 1) cupolas, 2) electric arc furnaces, and 3) electric induction furnaces.

Cupolas: Cupolas behave like small blast furnaces, and suffer many of the same problems and difficulties. Cupolas are the largest of the air pollution sources associated with foundries (Figure 7-3). The contaminants produced include metallic oxides, which are often very fine dusts, unburned hydrocarbons, and CO. There will also be some cinders. The traditional method of pollutant control is by use of a "wet cap" on top of the cupola (Figure 7-4). The gases are directed through a water curtain by their thermal impetus, and this water stream removes the larger particulates. Unfortunately, this method does remove only the larger particles, the resultant emissions still being quite contaminated. Figure 7-5 illustrates the results of a typical cupola emission control system. To meet the current emission standards, venturi scrubbers or fabric collectors, often in conjunction with mechanical separators, are used. Wet scrubbers can also be employed (Figure 7-5). Because of the large dust load, electrostatic precipitators are not usually economical. Fabric collectors are sometimes also impractical, since not only must the discharge be kept at a sufficiently cool temperature, but in colder climates the high-humidity dust could allow water condensation and clogging of the filter fabric.

Electric arc furnaces: Electric arc furnaces operate on the same principles as those in steel plants. Though significantly less dirty than cupolas, there are still

FIGURE 7-3. Top of cupolas with the capping off.

FIGURE 7-4. Schnieble cupola wet cap. (*Photo courtesy Wisconsin DNR.*)

FIGURE 7-5. Typical emissions of a wet cap collection device. (*Photo courtesy Wisconsin DNR.*)

FIGURE 7-6. Foundry after installation of wet scrubber system.

some air pollutants formed. Smoke is emitted from the furnace openings, particularly when the power is applied, and must be vented. The particulates from electric furnaces tend to be larger than those from cupolas, and thus are easier to collect. However, they also possess a characteristic color and appearance, and the finer particles readily scatter light, making them very visible. An extremely high collection efficiency is required to obtain a clear discharge. The exhaust gases are usually taken to fabric or wet collectors, or to electrostatic precipitators, for particulate removal.

Electric induction furnaces: These have been used for many years in nonferrous foundry operations, and they are now being used in iron foundries, too. Induction heating is based on the principle that an eddy current will be induced in any electric conductor located in a changing magnetic field. These eddy currents produce heat because of the normal I^2R losses experienced with any electric current, and temperatures to 3000°C are possible. In practice, the magnetic field is produced by a coil carrying an alternating current; this coil usually surrounds the substance to be heated. The eddy currents themselves can either be created directly in the substance, or they can form in a magnetic case surrounding the charge. In that latter case, the heat must be radiated inward to the charge.[2]

Induction heating does require large quantities of electric power. However, the melting method is clean, and normally no exhaust ventilation is required. During charging, fume emission may occur, and the furnaces must be hooded. All of these heating methods also produce slag, which must be disposed of.

Nonferrous Operations

The previous descriptions apply primarily to ferrous operations. The same principles apply, however, to nonferrous foundries. Nonferrous furnace operations have not, to date, been studied as much as the ferrous operations, so less is known about the nature of the effluents.

Copper-based alloys such as brass and bronze usually emit fumes of the minor metals present, not copper fumes. For example, the fumes from a brass furnace are primarily zinc oxide, with trace amounts of lead oxide. The collection equipment is thus designed for zinc oxide removal. The fumes emitted are white and very fine (usually less than 0.5 μm). In addition, the stack temperatures are usually higher than for ferrous operations, so fabric collectors cannot be used without precooling the dust.

No cases of community lead poisoning have ever been traced to foundry operations, in spite of the large amount of lead used in alloys and the high toxicity of lead. On the other hand, cases of beryllium poisoning have been reported when that substance has been used in copper alloys. The recommended atmospheric concentrations of beryllium are thus only 0.01 μg/m,[3] is a factor of 10 less than believed to be potentially harmful.

Alloys of aluminum and magnesium normally require fluxing agents to aid in the melting. These fluxing agents create the majority of the pollution, which consists primarily of either chlorides, fluorides, sulfur dioxides, or oxides of the alkali metals. Also present may be oxides of the materials comprising the melt. The gaseous fluorides are the most difficult of the pollutants to handle. Some wet scrubbers are effective in removing the fluorides, but corrosion is a problem. The effluent water must be chemically treated in many instances. Dry collectors may be used, if some finely divided, absorbent reagent is simultaneously introduced into the air stream to absorb the fluorides for later removal. Tall stacks may be used to some extent, where not restricted by government regulations.[3]

Pouring and Mold-cooling

The molten metal is transferred from the furnace to a pouring area by a transfer ladle. From the pouring area it can either be 1) poured directly into molds, or 2) transferred into a pouring ladle, from which it is then poured into the molds. The poured molds are allowed to partially cool, at least until the metal has solidified, and then the castings are removed from the mold, rough cleaned, and allowed to

FIGURE 7-7.   A series of sand molds and cores to be used to produce a variety of decorative objects.

further cool until there is no difficulty in handling them. During the pouring operation, smoke is emitted, and thus the area must be hooded and vented. An amount of fines from the molds is released at this stage. These fines are also caught by the venting system.

Molding Department

The molds are used to form the outside shape of the castings. These molds are made of a mixture of sand and various other substances that solidify into almost any shape desired. For the "gray" and ductile iron foundries, for example, they typically contain silica sand, western bentonite (clay), seacoal (carbon), cellulose, and water, and are called "green sand." The vast majority (95% or better) of this green sand can be reused. To be able to do so, the molds, once removed, must be further broken down in a mixer. This is a very dusty crushing and grinding procedure, and the fines must be collected by wet scrubbing, or possibly by dry mechanical means, but that latter process may be inadequate. Fabric collectors generally are not used, for the fabric may clog due to moisture condensation from the air stream. The approximately 5% that cannot be recycled has been thermally degraded and no longer contains the required molding characteristics. This waste green sand must be landfilled at present. Technology is available to recover the silica sand from the green sand, but this is economically prohibitive considering today's energy costs and the abundance of natural deposits of silica sand in many locations in the United States.

Cast-cleaning Department

The cast-cleaning procedure consists of three types of steps. The castings, after their removal from the molds, must first have the waste metal knocked off as much as possible. This waste metal is remelted and reused. A dust exhaust system is required.

The castings are then further cleaned by grinding. This procedure throws dust in all directions, so the area must be thoroughly hooded.

Lastly, the castings are subjected to an abrasive blasting. Some type of abrasive material, such as steel shot, is impinged against the castings with large volumes of air. This air is usually of low humidity, so fabric collectors can be used to collect the resultant particulates. Wet scrubbers are also feasible except in cases where steel shot is used: in those cases, a buildup of Fe_2O_3 will occur at all points of direction change of the air stream.

Core Room

The core is the interior of the mold, necessary to shape a hollow casting. The cores are similar in chemical composition to the molds, but usually contain less moisture, and thus bag collectors can be used for particulate collection. Binders are frequently used to form the cores, substances such as furfurol (a product derived from corn). Binders can decompose, resulting in the emission of organic sulfur compounds, phenols, aldehydes, unsaturated hydrocarbons, and other organic compounds, many that are odorous. Plants may just dilute these odors, or they may need to employ wet scrubbing procedures for their removal. The reprocessing of the cores is similar to that of the molds.

Pattern Shop

In order to form a mold or core, it is necessary to start with a "pattern." This pattern is essentially identical to the finished casting desired, except that usually it is constructed of wood, for ease of production. The pattern is surrounded by green sand, which is then allowed to harden. After solidification of the sand, the pattern is removed, leaving the "finished" mold, which now has the proper interior configuration to replicate (now in metal) the pattern. A foundry will typically have hundreds of patterns stored, available to produce molds for all of the requested castings. Sawdust is a common waste product in the pattern shop.

Maintenance Shop, Casting Repair Areas

There are typically two areas of a foundry that are devoted to repair work: the maintenance shop, in which all necessary equipment repairs are made, and the

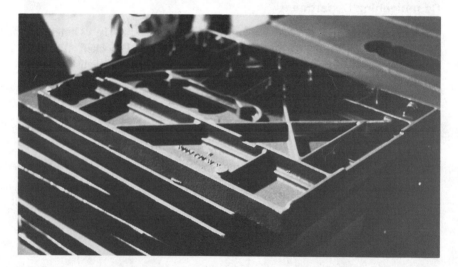

FIGURE 7-8.   Typical metal pattern which is a replica of the final item, and is used to produce the molds and cores.

casting repair shop. In both of these areas, welding and soldering create smoke and fumes, which must be vented.

BYPRODUCT RECOVERY

The dusts and fumes collected in the ferrous foundry industry are essentially worthless; some of those collected in the nonferrous operations may have some value, perhaps enough to pay for the handling of them after collection. In several cases, the Fe_2O_3 particulates have been collected from the electric furnace operations, bagged, and then fed back into the furnace to aid in carbon control. Overall, however, the air pollution equipment is essentially a nonproductive expense, and additional costs are incurred for solid waste disposal.[4]

OTHER POLLUTANTS

Water pollution problems arise mainly from efforts to clean the emitted air. Tons of very fine airborne solids are collected daily by wet scrubbing. Some phenols (from the binders), acids, lubricating oils, and cleaning wastes are also possible. If so, this waste water may need to undergo some oxidation procedure before disposal.

The wastewater is typically treated by either of two ways: separation processes or transformation processes. Transformations from one type of substance to another are sometimes necessary to allow effective separation, and are always followed by a separation of some type.

Several specific processes that can be used for separation are the following:

1. Sedimentation that can occur in lagoons if sufficient land is available. The retention time is measured in days (up to 30 days is desirable), so a very large surface area is required.
2. Detaining the water, for a short term, to allow the heavier solids to settle, and then flocculating and sending it through mechanical clarifiers with sludge collection equipment, after which the sludge can typically be dewatered by a vacuum filter.
3. Flotation, which can be used for the lighter-than-water particles and those occurring as a finely divided phase. It can also be used to thicken the concentrated wastes and sludge.
4. Filtration, via screens, for the larger-size particulates, those 1/4-in or larger. Some success has been obtained for particles as small as 10μm in diameter.
5. Advanced treatment methods, such as reverse osmosis, activated carbon absorption, and ion exchange. Though not generally used to treat foundry wastes, these can be employed when it is necessary to obtain a very pure water for specialized uses.

The transformations that can occur as part of the treatment process can be by either chemical or biochemical techniques. Chemical treatment typically involves precipitation, coagulation, or oxidation. Biochemical treatment usually involves oxidation.

Chemical precipitation involves the addition of appropriate counter-ions to form insoluble precipitates. These precipitates, containing the ions whose removal is desired, are separated from the wastewater by one of the normal separation techniques, such as sedimentation or flotation. Chemical precipitation is most widely used for removal of ions of metals such as iron, copper, zinc, or chromium.

Chemical coagulation involves the destabilization and aggregation of dispersed materials, followed again by separation. Most commonly, salts that yield trivalent aluminium or ferric ions are used, but various organic polymers are gaining in popularity.

Chemical oxidation gasifies, precipitates, or otherwise reacts with the pollutants to make them less of a waste problem. Chemical oxidation is, however, not as frequently used as biochemical oxidation, which is almost as effective and is less costly. This biochemical oxidation can be done as part of the lagoon sedimentation process, or in activated sludge or trickling filter-type plants, depending upon the particular foundry characteristics.[5]

The scrubber sludge, combined with the slag created in the melting operation, typically accounts for 15–20% of the landfilled refuse, the remainder being green sand. There is little use for either the sludge or the slag: some lightweight building blocks utilize a small amount of slag, but the market for it is presently very poor. Table 7-1 lists the metals present in typical foundry sludge (ferrous operation). High heavy metals concentrations must be monitored carefully before disposal, for some

TABLE 7-1   Elemental Metals Composition of Typical Foundry Scrubber Sludge (Values accurate within a factor of 2; i.e., a reported 30% may be between 15% and 60%.)

Substance	Percent
Ca	30.0
Fe	10.0
Si	10.0
Mn	3.0
Mg	3.0
Al	3.0
Zn	1.0
K	1.0
Pb	0.8
Na	0.7
Sn	0.1

TABLE 7-1.   *(Continued)*

Substance	Percent
Ba	0.05
Cu	0.05
Cr	0.03
Ti	0.03
Sr	0.02
Cd	0.01
Ag	0.008
B	0.005
Ni	0.005
Zr	0.005
Ir	0.005
Mb	0.001
Bi	0.001
V	0.0007

may be toxic. However, toxicity is generally not a problem with normal ferrous operations.

References

1. *Control of External Air Pollution.* American Foundrymen's Society, Inc., 1976, p. 3.
2. Perry, John *et al., Perry's Chemical Engineers' Handbook, 4th Edition.* New York: McGraw-Hill, 1969, pp. 25–42.
3. *Control of External Air Pollution.* p. 9.
4. *Control of External Air Pollution.* p. 13.
5. *Water Pollution Control.* American Foundrymen's Society, Inc., 1974, pp. 13–25.

8

Nonferrous Metals Production

Many nonferrous metals such as copper, zinc, lead, and aluminum are extracted from the ores and purified in smelter operations. This chapter will consider specifically two of these metals: copper and aluminum.

The extraction of all metals creates hazardous waste management problems worldwide. For example, beneficiation and smelting release an estimated 7–70×10^3 metric tons of metals to the aquatic environment. The ores frequently contain large percentages of sulfide ($>30\%$), with pyrite (FeS_2) being the primary component. Other contaminants include antimony, arsenic, cadmium, lead, zinc, and sometimes copper, itself.[1]

COPPER

Copper Production

Copper ore occurs naturally, mainly as the sulfide, Cu_2S, called chalcocite, with smaller amounts of copper carbonates and copper pyrites also present. Some deposits of copper occur instead as copper oxides. Copper is often found in conjunction with lead, zinc, silver, and gold. An overall purification process is illustrated in Figure 8-2.

Mining and Concentration
Copper is mined primarily by open pit mines. (Figure 8-3). Concentrations of copper as low as 0.5% are currently economical to extract. Approximately two tons of overburden must be removed with each ton of ore, on the average; the ratio can be as large as 5 to 1. Typically, the ore is concentrated near the mine, and the concentrate then is shipped to other locations for smelting and refining, similar to the procedure used for iron production.

The concentration (beneficiation) process consists of four steps: crushing,

162

FIGURE 8-1.   The White Pine Copper plant in Upper Michigan, near Lake Superior. Shown are the smelter feed and bedding building, smelter building, and the 500-ft tall stack.

grinding, oil flotation, and dewatering. The large ore chunks are first crushed to less than 1 in diameter (Figure 8-4), and then ground finely by rod or ball mills (Figure 8-5). The Cu_2S grains can then be separated from the tailings by an oil flotation procedure (Figures 8-6 to 8-8). In this process the finely ground particles are treated with a mixture of oil and water. The Cu_2S is wet by the oil, and floats in the upper oil layer; the silicates are wet by the water, and thus sink to the bottom through the lower water layer. Lime and/or fuel oil can be added to minimize the effects of the bitumen found in the ore. The Cu_2S is removed by aerating the oil to form a froth. Pine oil and/or an alcohol may be added as frothers. The froth is then collected and dewatered by settling and vacuum filtration, and the resulting impure Cu_2S (up to 50% copper) is shipped to the smelters.

For every 1000 tons of ore mined, approximately 160 tons of concentrate and 840 tons of tailings are formed. These tailings typically consist of 50–70% quartz,

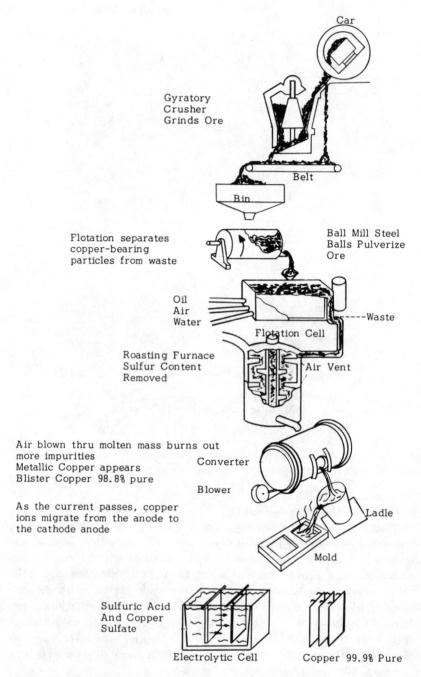

Car

Gyratory
Crusher
Grinds Ore

Belt

Bin

Flotation separates
copper-bearing
particles from waste

Ball Mill Steel
Balls Pulverize
Ore

Oil
Air
Water

----Waste

Flotation Cell

Roasting Furnace
Sulfur Content
Removed

Air Vent

Air blown thru molten mass burns out
more impurities
Metallic Copper appears
Blister Copper 98.8% pure

Converter

Blower

As the current passes, copper
ions migrate from the anode to
the cathode anode

Ladle

Mold

Sulfuric Acid
And Copper
Sulfate

Electrolytic Cell

Copper 99.9% Pure

FIGURE 8-2. A traditional copper purification process. Newest technology combines the roasting and converting steps, or replaces them with a leaching process.

FIGURE 8-3.   Mining pit at Cananea in Sonora, Mexico. This area had, at the time of the photograph, approximately six months of production remaining. (*Photo courtesy Mountain State Mineral Enterprises, Inc.*)

micas, and clays, and 30–50% iron sulfides (pyrites). These tailings are recovered mixed with water from the beneficiation and are often pumped to a "tailing pond" for settling. The excess water can be pumped back to the concentrator plant, where, for every 1000 tons of ore treated, about 400,000 gallons of water are required.[2] These tailings are fairly reactive and quite readily form sulfuric acid (H_2SO_4), as do some iron ore tailings. Frequently, they, too, must be appropriately handled and treated to minimize the environmental effects (see Chapter 6).

In addition to tailings, some "slime" is formed during the flotation. This slime can be in amounts up to 37% of the weight of the feed, and must be discarded.

Smelting
After flotation, the ore is reduced to the metallic state, most often by a pyrometallurgical (smelting) process. Traditionally, the concentrated ore has been processed in a reverberatory furnace. The chemical composition of a typical copper concen-

FIGURE 8-4. Looking from adjacent to primary crushing structure across secondary/tertiary crushing facilities to grinding facilities. (*Photo courtesy Mountain States Mineral Enterprises, Inc.*)

FIGURE 8-5. Primary grinding area of Copper Range Co. concentrating plant showing the 10 ft. long, 13 ft. diameter ball mills and 84-inch diameter screw classifiers. (*Photo courtesy of Copper Range Co.*)

FIGURE 8-6. Tapping a reverbatory furnace in the smelter. (*Photo courtesy of Copper Range Co.*)

FIGURE 8-7. Closeup of the flotation process.

FIGURE 8-8.   The 400-ft thickener building adjacent to the grinding facilities at Cananea, Sonora, Mexico. (*Photo courtesy Mountain States Mineral Enterprises, Inc.*)

trate, as sent to the reverberatory furnace, is in Table 8-1. Initially, the FeS oxidizes: $2FeS + 3O_2 \rightarrow 2SO_2 + 2FeO$. The FeO then reacts with the SiO_2, forming slag: $FeO + SiO_2 \rightarrow FeSiO_3$ (slag). This removes many of the iron and silica impurities, and leaves behind copper "matte," containing primarily Cu_2S and FeS. Reverberatory technology is rapidly being replaced by oxygen/flash smelting, which dramatically reduces the volume of off-gases.[3]

TABLE 8-1   Chemical Composition of Copper Concentrate[a]

Substance	Composition (%)
Moisture	20
Copper	35
Al_2O_3	8
Sulfur	7
Iron	5
CaO	2
MgO	3
SiO_2	28

[a]Data from Ramsey, R. H., "White Pine Copper." *Engineering and Mining Journal*, January 1953, p. 85.

FIGURE 8-9.   Converter aisle looking west at Cananea. (*Photo courtesy Mountain States Mineral Enterprises, Inc.*)

Further processing can be done in a converter, similar in principle to a Bessemer converter. If the concentrate does not contain adequate iron or CaO, additional must be added to form a good quality slag. Sand is added to the matte, and hot air is blown through the molten mixture. The iron present reacts with the sand, and forms more slag, $FeSiO_3$, and the CuS releases SO_2: $Cu_2S + O_2 \rightarrow 2Cu + SO_2$. The solid copper (99% pure) is called "blister" copper due to the bubbles formed by the escaping SO_2. In addition, the CuO that is present is also converted to copper at this stage.

The most recent technology combines this converter step with the smelting step.

Leaching

Hydrometallurgical processing, or leaching, is becoming an increasingly important alternative to smelting, particularly for non-sulfide ores (oxides, silicates, or carbonates). Weak acid is percolated through ore or rejected materials. The copper

FIGURE 8-10.   Converter pouring and sampling in process.

is leached out and then recovered from the leach liquor, frequently by solvent-extraction (Figure 8-11).[4,5]

The leachate enters the extraction area, a series of mixer-settlers, where it is contacted by a countercurrent flow of organic extractant. This extractant consists of a number of special organic compounds dissolved in a kerosene carrier fluid. The percentage of copper extracted is typically >95%, but depends upon the concentration of copper and acid in the leachate and the number of contacting stages.

The copper-containing organic solution is then contacted with the spent electrolyte from the electrolysis step, frequently called the "electro-winning" process, in a separate series of mixer-settlers. Additional makeup acid must then be added to the electrolyte, and the copper can then be further purified by electrolysis.

The major advantage of a solvent-extraction procedure is the reduction of air pollution, as compared to smelter operations. The only effluents, in fact, are the spent acid remaining from the extraction, which can be recycled to the leaching, and a small amount of electrolyte, which must be bled to reduce the iron concentration. The leaching procedure dissolves only trace amounts of iron, but without this bleed, the iron would concentrate in the electrolysis operation, lower its efficiency, and lead to potential corrosion.

This bleed solution can also be recycled to the leachate. A small amount of the

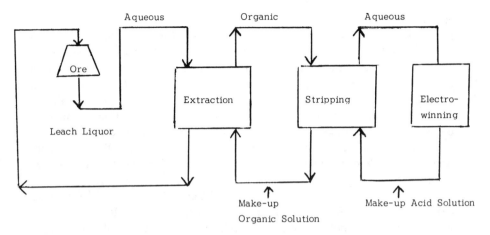

Aqueous   Organic   Aqueous

Ore

Leach Liquor

Extraction   Stripping   Electro-winning

Make-up
Organic Solution

Make-up Acid Solution

FIGURE 8-11.   Schematic of the solvent-extraction operation.

organic solution will be lost to the leachate in the contact process ($\sim$.01%), but much of this can be collected by flotation, if desired.

A major limitation of the leaching process is that sulfide ores cannot be effectively leached by acidic solutions; sulfide ores must be solubilized by use of bacteria. In some locations, this bacteriologically promoted leaching occurs naturally, and attempts are being made by some researchers to eventually control this natural process. At minimum, additional aeration can be provided.

Fire Refining

Blister copper, from the converter or leaching operations, can be purified to $<$ ½% impurities by a refining process. Some of these impurities can best be removed by "fire refining," which is conducted in small reverberatory furnaces or revolving furnaces.

These furnaces are equipped with fuel burners that melt the blister copper and maintain it in the molten state. Simultaneously, air is forced through the material to oxidize the remaining impurities. The impurity oxides float to the surface, from where they are skimmed.

Subsequently, reduction is carried out by introducing green (fresh) logs into the molten pool in a process called "poling." The green logs emit highly reducing gases that can remove the oxygen from most of the copper oxide present in the melt. An alternative would be to bubble natural gas through the molten material to reduce the oxides.

The resulting metal is called tough-pitch copper. This is a commercial product, though it does contain some gold, silver, and other impurities. Most of the tough-pitch product is, therefore, cast into anodes for further refining.

Electrolysis

Electrolysis is accomplished using a copper sulfate ($CuSO_4$) solution. Two electrodes are placed in a tank containing this solution: pure copper is used as the cathode, and the crude, tough-pitch copper as the anode. A voltage is applied across these two electrodes. The half reactions that occur are $Cu \rightarrow Cu^{2+} + 2e^-$ at the crude copper anode, and $Cu^{2+} + 2e^- \rightarrow Cu$ at the pure copper cathode (Figure 8-12). The net effect is that impure copper is dissolved from one electrode and pure copper plates out on the other electrode. The separation of the copper from the impurities occurs because of their differing propensities for oxidation and reduction. Iron has a higher oxidation potential than does copper; therefore, it dissolves readily to form the Fe^{2+} ion. However, reduction (back to Fe) is more difficult, so the Fe^{2+} remains in solution. On the other hand, gold and silver have lower oxidation potentials than does copper—low enough so the metals never become ionized, and instead fall unoxidized to the sludge at the bottom of the tank. These metals, of course, are valuable byproducts, and are subsequently isolated by another process, often by use of a roaster.

Pollution Problems Associated with Purification

The major environmental concerns associated with the purification steps are two-fold: 1) the SO_2 produced in the smelting is very harmful to humans and vegetation, and 2) production of the large quantities of power required for electrolysis forms additional SO_2 plus several other types of environmental pollutants (Fig. 8-13).

To obtain an idea of the magnitude of the SO_2 problem, a 1969 study of the copper smelters west of the Mississippi River[6] found that approximately 1,565,000 tons of SO_2 were generated annually, and about 284,000 tons, or 18.1% of the SO_2, were recovered. This SO_2 can create a very hazardous situation. The Copper Hill,

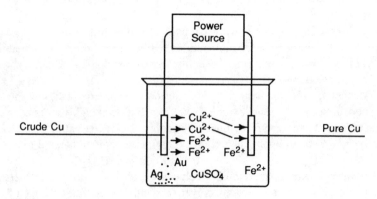

FIGURE 8-12.   Schematic of the electrolysis procedure.

FIGURE 8-13. The SO_2 emitted from copper and other non-ferrous metal processing has devastated many acres around smelting operation such as this area near Sudbury, Ontario.

Tennessee smelter, for example, is situated in the center of a 25-mile square area virtually devoid of vegetation. This area has been almost impossible to rehabilitate even after 40 years.[7] Recovery of the SO_2 is difficult because of the low concentrations found due to air dilution. The recovery costs of the SO_2 are inversely proportional to its concentration in the air stream; hence, it would be preferable to prevent this dilution, if possible. The new oxygen flash smelting process does produce a more concentrated gas, and at those sites sulfuric acid is an important by product.

Though H_2SO_4 recovery is most common, there are three processes that have been used, at least to some extent, to recover the SO_2 generated in the smelting operations. These consist of 1) H_2SO_4 manufacture, 2) the production of elemental sulfur, and 3) the concentration of weak sulfur oxides by absorption.

The SO_2 can be converted to marketable H_2SO_4 (93% or more) by a contact process. The gas is usually first cleaned by an electrostatic precipitator, and then further processed in four operations for sulfur trioxide (SO_3) removal:

1. A humidifying tower
2. A cooling tower
3. A gas cooler
4. An electrostatic mist precipitator.

The gas from the furnaces usually contains both SO_2 and SO_3. In the humidifying tower, water vapor is introduced into the gas. The SO_3 present rapidly reacts with the water to form H_2SO_4: $H_2O + SO_3 \rightarrow H_2SO_4$. This acid is cooled and somewhat dried in the cooling tower and gas cooler, and collected in the mist precipitator. The H_2SO_4 produced in this stage is very weak and often not further processed. It can be used for leaching in the smelter, or can be neutralized and wasted.

After the removal of the SO_3, the gas that now contains SO_2 and H_2O is further dried by direct contact with very strong (93%) H_2SO_4 in a packed tower arrangement. The concentrated H_2SO_4 in the tower absorbs much of the water from the gas and itself becomes more dilute (and thus must be replaced continually) and is added to the product system.

The dried gas, containing SO_2, is then heated and allowed to flow to a catalytic converter for conversion of the SO_2 to SO_3. The gas stream next flows through an absorption tower, where the SO_3 is absorbed in strong H_2SO_4. As previously, the SO_3 reacts with any water present to generate H_2SO_4. Now, however, the net acid is extremely strong. Some H_2SO_4 remains in the gas stream and must be removed. This last step again produces a dilute H_2SO_4, which is often wasted.

Overall, the efficiency is quite high, but this depends upon the initial SO_2 concentration. Typically, the recovery is about 93% for an 8% SO_2 gas, but only 89% for a 2% gas.

Elemental sulfur can be produced by several reduction processes. One process, developed by the American Smelting and Refining Company, requires sufficient natural gas to react with all the O_2 present in the waste gas plus that required to reduce the various sulfur oxides to sulfur. The methane (natural gas, CH_4) is mixed with the hot waste gas as it enters a brick-lined combustion chamber. The resulting reaction between the CH_4 and SO_2 reduces the SO_2 to sulfur: $CH_4(g) + 2\ SO_2(g) \rightarrow 2S(g) + 2H_2O(g)$. In addition, several side reactions produce hydrogen sulfide (H_2S) and carbonyl sulfide (COS). After partial cooling of the gas, the COS can be converted to sulfur by a bauxite catalyst. Further cooling liquifies the sulfur, which can be collected by a precipitator. The H_2S remaining in the gas stream can then be converted to sulfur by a further catalytic procedure; the gas is reheated, subjected to two catalysts in series, cooled, and liquid sulfur is again collected by a precipitator. The total sulfur recovery is about 95%. The necessity for CH_4 periodically limits use of this procedure. Not only is it expensive, but in some localities natural gas is at times in short supply.

Another method for production of elemental sulfur is by direct reduction. One technique calls for reduction by H_2 in the presence of a suitable catalyst. The first stage converts SO_2 to H_2S; this H_2S is then reacted with more SO_2 to form sulfur. This technique may be feasible for waste gases with high concentrations of SO_2 and low concentrations of O_2.

SO_2 can also be reduced directly by coke. Several processes have been developed, and all involve passing the hot waste gas through a bed of coke. Various other reaction stages are also necessary to control side reactions, such as the production of H_2S.

An absorption process can be used, often in conjunction with a contact H_2SO_4 plant or a reduction process, to concentrate the SO_2 from, for example, 0.5% to 24%. As previously noted, the dilute gases are less economical to treat, so preconcentration is often beneficial. One particular process involves the scrubbing of the gas by a solution of ammonium sulfite, $(NH_4)_2SO_3$, and ammonium bisulfite, NH_4HSO_3, salts. This solution is maintained at a low pH (acidic) and high salt concentration. The SO_2-containing waste gas is passed through this scrubber solution, is absorbed, and reacts with the sulfite to form the bisulfite: $SO_2 + SO_3^{2-} + H_2O \rightarrow 2HSO_3$. The gas is then taken to a second scrubber for further SO_2 removal, this time by a high-pH solution with a low salt concentration: $SO_2 + OH^- \rightarrow HSO_3^-$. The waste gas is then released to the air, while the bisulfate solutions are acidified with H_2SO_4, and then stripped of the resulting SO_2 by an air stream: $HSO_2^- + H^+ \rightarrow SO_2 + H_2O$. This SO_2 in the stripping air is concentrated enough for economical H_2SO_4 recovery.

The difficulties associated with the large power consumption are particularly experienced with aluminum production, and thus will be discussed within that section, rather than here.

Clark Fork Complex[1]

The Clark Fork River Basin, in western Montana, has been identified by the EPA as the largest complex of Superfund sites. This area, encompassing an acreage 1/5 the size of Rhode Island, was contaminated by > 125 years of copper and silver mining and smelting. The wastes generated consist of tailings, slag, flue dust, and waste rock, which was initially deposited near the sites. As the pollutants were transported by rivers or the wind, they produced more widespread contamination. Subsequent remobilization of these distributed wastes has led to additional environmental problems.

Very different heavy metals predominate in the various types of waste, as seen in Table 8-2.

The waste rock generally has little contamination, though it can cover a significant amount of land. At the Clark Fork Complex, 300 million m^3 of rock

TABLE 8-2   Metal Concentration (mg/kg) in Various Waste Deposits[a]

Metal	Slag	Tailings	Flue Dust
As	1070	2960	10,400
Cd	13.4	8.0	—
Cu	7000	6730	37,100
Pb	1030	2740	—
Zn	18,000	11,000	—

[a]Data from Moore, Johnnie N. and Samuel N. Luoma, "Hazardous Wastes from Large-Scale Metal Extraction: A case Study." ES&T, **24**, *No. 9*, September 1990, p. 1280.

were removed from the Berkeley open pit mine and tens of millions of cubic meters were removed from the underground mines. This waste rock was spread over approximately 10 km² (3.9 square miles) of land.

Within the Clark Fork deposits, approximately 90% of the ore-formed tailings. The tailings ponds cover > 35 km² (14 square miles), and contain > 200 million m³ of tailings. Based on the concentrations in Table 8-2, this means that the ponds probably contain 9000 metric tons of arsenic, 200 metric tons of cadmium, 90,000 metric tons of copper, 20,000 metric tons of lead, 200 metric tons of silver, and 50,000 metric tons of zinc. The presence of large amounts of organic matter in the ponds has produced anaerobic conditions, which has immobilized the cadmium, copper, lead, and zinc as sulfides. Arsenic, however, is relatively free and could readily move into groundwater.

Early smelting was conducted primarily by "heap roasting," or intermixing ore and timbers, and then burning the wood. This process, common in the late 1800s and early 1900s, released massive amounts of SO_2 and metals into the atmosphere, and poisoned countless cattle, sheep, and horses. Even today, soil contamination visibly affects an area of at least 300 km² (120 square miles) surrounding the Anaconda smelter site.

During the underground and surface mining operations in the Clark Fork Complex, acidic metal-contaminated groundwater was pumped from the mines. After the mining was discontinued in 1982, the groundwater was no longer pumped away, and the 390 m deep Berkeley pit and the various underground shafts and tunnels began filling. This water is highly contaminated with metals and sulfates. Estimates are that the pit contains 70 metric tons of arsenic, 20,000 metric tons each of copper and zinc, and 100,000 metric tons of sulfur. Since groundwater flow raises the water level in the pit about 22 m/y, it is estimated that the pit will overflow near the year 2000, spreading the contamination into an adjacent aquifer or perhaps onto the ground surface.

Until the early 1900s, much of the waste materials from beneficiation and smelting was sluiced onto surrounding land or into streams. These huge quantities of sediment plugged stream beds, leading to flooding and deposition of contaminants on surrounding land. Even today, bright blue and green copper sulfates can be seen on banks up to 200 km downstream; areas where no plants have survived are found over 100 km from the smelting and beneficiation sites.

Remobilization of these contaminants has been noted > 200 km from the mines and smelters at Butte and Anaconda. The Milltown Reservoir retains about 100 metric tons of cadmium, 1600 metric tons each of arsenic and lead, 13,000 metric tons of copper, and 25,000 metric tons of zinc. Nearby wells were contaminated with excessive arsenic and have had to be closed by the EPA.

The overall environmental effects of the mining and smelting have been diverse. Among other measured effects, trout densities in the Clark Fork are only about 2% of what would otherwise be expected in a similar Montana river. Studies comparing

3000 U.S. counties between 1959 and 1970 place Silver Bow County, surrounding the Butte mining area, among the 100 counties with the highest mortality rates for people aged 35 to 74. National Cancer Institute comparisons of cancer rates also show increased incidence of some cancers for both men and women in the Clark Fork Basin.

Successful remediation of this site will require a wide variety of techniques, which can be identified only after extensive mapping of contaminant flows, analyzing of medical and chemical data, and undoubtedly the development of new, specific, and innovative solutions. It is not expected that the area will ever be restored to its pre-industrial status.

Copper Reclamation[8]

Proper recovery of scrap copper, as well as other scrap metals, can greatly reduce the energy demands and lessen the pollution of our environment. In addition, many nonfuel resources, such as the metal ores, may also become scarce in the immediate future. Even now, the United States has very few of its own supplies of many critical resources. To reduce our dependence upon foreign governments and to help our balance of payments and thus maintain the value of the dollar, it is essential that we do recycle as much as possible of these valuable resources. Unfortunately, the recovery of these metals is not without environmental difficulties.

A large portion of the reclaimed copper is from scrap copper wire of all sizes—from large utility power cables to common household wire. In the majority of cases, this wire must be removed (by hand) from some piece of electrical equipment. Frequently, an insulator coating must also be removed from the wire. This is the major reclamation difficulty. There are two methods commonly used to remove this coating: 1) the coating can be mechanically stripped if the cable is large, and 2) the insulation can be burned off under controlled conditions. Other mechanical separation methods, such as pulverizing the complete wire in a hammer mill almost to powder, and then separating the copper and insulation materials by utilizing their differing densities, are also being developed.

In recent years, more than half of the copper consumed in the United States has been recycled scrap. Of this scrap, 50-60% is considered "new" scrap, that is, recovered from copper machining operations. The remainder, the "old" scrap, is from recycled consumer products (electrical cable, automobile radiators). Internal plant recycle is not reflected in these numbers.[3]

Environmental Problems
Air pollution from incineration of the insulating jacket is the major environmental threat of copper recycling. The coatings can be classed into three categories: 1) halogenated plastics; 2) nonhalogenated plastics, cotton, paper, silk, and rubber; and 3) metallics. The metallics cannot be combusted, so they must be mechanically

stripped. The nonhalogenated plastics, paper, fabrics, and rubber can theoretically be completely combusted, producing only CO_2 and H_2O. The required combustion conditions necessitate a two-stage process. The first ignition is in a primary combustion chamber operating at 1400–600°F for 0.3–0.6 seconds. This must be followed by treatment in a secondary chamber operating at > 1400°F for > 0.7 seconds, with good mixing of the remaining substances. The halogenated plastics are the most troublesome to incinerate, since they result in a variety of potentially harmful emissions.

These halogenated plastics are typically substances such as polyvinyl chloride (PVC) or teflon. When incinerated, they produce large amounts of hydrochloric acid (HCl) and hydrofluoric acid (HF), respectively. For example, the burning of PVC has reportedly led to the emission of HCl equivalent to 58% of the weight of the PVC sample. In addition, incomplete combustion leads to the emission of some photochemically reactive and/or odorous organics, including aromatic, aliphatic, and olefinic hydrocarbons.

The air pollution control systems for these gases must consist of two stages: 1) an afterburner to eliminate the carbonaceous particulates and hydrocarbon gases, and 2) a caustic wet scrubber to remove the residual HCl or HF.

ALUMINUM

Aluminum Production

Aluminum is an active, easily air-oxidized metal. Normally, the metal surface attains a thin oxide layer that prevents further reactions and thus makes it very durable. Aluminum is particularly attractive for many purposes also because of its low density, only 2.7 times that of water (compared to 8.9 for copper and 7.9 for iron or steel).

Aluminum is a relatively abundant ore (in fact, aluminum ranks in abundance behind only oxygen and silicon in the earth's crust), occurring primarily as bauxite, $Al_2O_3 \cdot xH_2O$.

The overall aluminum production process is illustrated in Figure 8-14.

Purification

The bauxite now mined is of sufficiently high grade so that generally no beneficiation is required at the mine. At the smelter, the aluminum ore is first crushed, cleaned of impurities, and then dissolved in sodium hydroxide (NaOH), a strong base: $Al_2O_2 \cdot xH_2O + 2OH^- \rightarrow 2AlO_2^- + (x + 1)H_2O$. Aluminum oxide ($Al_2O_3$) is one of the few metals that will dissolve in bases as well as in acids, so the metallic impurities tend to precipitate out. Iron, for instance, is one of the primary impurities, and it precipitates as the hydroxide, $Fe(OH)_3$, and as its silicates, Fe_2SiO_4, $FeSiO_3$, and FeSi.

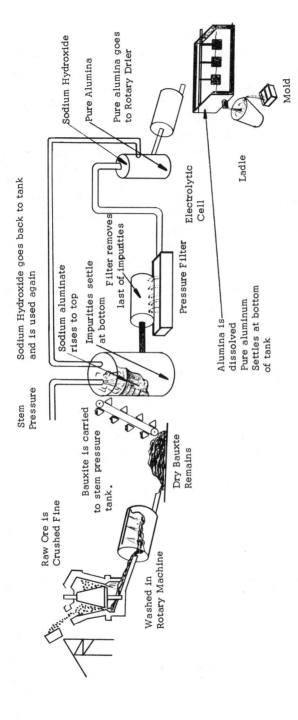

FIGURE 8-14. The overall aluminum production process.

Next, the basic solution is filtered to remove all the precipitated impurities. The filtrate containing the AlO_2^- is then treated with CO_2 to lower the pH. When the mixture is thus acidified, the AlO_2^- reacts to form aluminum hydroxide, $Al(OH)_3$, which precipitates and is subsequently removed.

The $Al(OH)_3$ is heated strongly (to about 1000°C) in a reverberatory furnace (Fig. 8-15) or in a small crucible furnace: $2Al(OH)_3 \rightarrow Al_2O_3 + 3H_2O$. The resultant Al_2O_3 is then further purified by electrolysis.

The Al_2O_3 is dissolved in molten cryolite (Na_3AlF_6) at high temperatures. Because of the required high temperatures, the electrolysis cell must be graphite lined. A carbon electrode is used as the anode, and the graphite lining can act as the cathode. The molten aluminum that is generated at the cathode, the lining, sinks to the bottom as it is formed because it is denser than the cyolite. The resulting half-reactions are:

$$\text{Cathode } Al^{3+} + 3e^- \rightarrow Al$$
$$\text{Anode } C + 2O^{2-} \rightarrow CO_2 + 4e^-.$$

After purification, the aluminum is usually pressed to form thin sheets (Figure 8-16) before formation into the final product.

Environmental Problems due to the Processing

There are several pollutants that are generated by the aluminum processing itself. One air pollutant generated during aluminum production, at least to some extent, is fluoride. It is inevitable that gaseous fluoride be generated from the molten fluoride salts due to the electrolytic processing. EPA regulations currently limit fluoride emissions to 1.9 lb fluoride per ton of aluminum.

One method for gaseous fluoride control, employed by the Aluminum Company of America (Alcoa), is by use of a fluidized-bed equipped with a baghouse. The

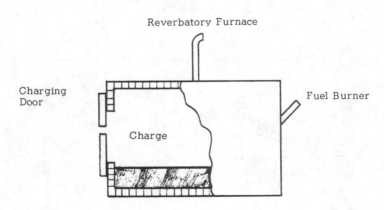

FIGURE 8-15.   Reverberatory furnace.

FIGURE 8-16.   Aluminum sheet being forced through press.

gaseous fluorides are adsorbed in the fluid bed; the particulate fluorides are trapped in the baghouse.[9,10]

Another waste is the "red mud" tailings produced from aluminum ores (Figure 8-17). This red mud is typically comprised of 20% Al_2O_3, 49% Fe_2O_3, with smaller amounts of silicon, calcium, sodium, titanium, phosphorus, and sulfur. Studies are being made as to possible byproducts and include use as thermal insulation, an additive to portland cement, concrete, or slag wool, and as a raw material for production of porous structural blocks. A process is currently being marketed for converting the red mud to steel by means of a moving grate. This grate dries, pelletizes, prereduces, and smelts the material by a basic oxygen process.[11] There is, however, no consistent market, as of yet, to consume the quantity of red mud produced annually.

Power Generation

As mentioned earlier, the major environmental difficulty associated with aluminum manufacture is the tremendous electrical power requirement for purification. It is estimated that about 10 kWh are required to produce a pound of aluminum from the ore, more energy than required by any other industry. Industry overall uses about 40% of the total U.S. energy consumption, and aluminum manufacture requires 10% of all the industrial energy. Table 8-3 illustrates the relative amounts of energy required for steel and aluminum.

FIGURE 8-17.   Aluminum sludge. (*Photo courtesy Wisconsin DNR.*)

Table 8-4 summarizes the sources of electrical power in the United States in 1990.

Many aluminum producers own their own electric power generating facilities (Figure 8-18), or own at least the fuel supplies (for example, the coal fields). Some operate their own hydroelectric generating plants. One aluminum company receives 45% of its energy from its own hydroelectric facilities. This company additionally owns coal fields or coal-fired steam power plants sufficient to provide one-half of the remaining 55% of its required energy.

Not only is the energy consumption in itself a major difficulty (even more so when one remembers that conversion to electrical power is only about 30% effi-

TABLE 8-3   Energy Required to Produce Metals From Ore or Recycled Metals[a]

Product	Energy (kWh/ton)
Steel plate or wire from ore	2700
Steel plate or wire from scrap	700
Aluminum from ore	17,000
Aluminum from scrap	850

[a]All data except for recycled aluminum is from Pyle, James L., *Chemistry and the Technological Backlash*. Englewood Cliffs, New Jersey: Prentice-Hall, 1974.

TABLE 8-4   1990 U.S. Electrical Power Generation[a]

Source	Million Kwh	Percent
Coal	1,557,498	55.5
Nuclear	576,784	20.6
Natural gas	263,452	9.4
Hydro	279,893	10.0
Oil	117,062	4.2
Other	10,645	0.4

[a]Data from *Info* 262, U. S. Council for Energy Awareness, March 1991, p. 1.

cient, requiring tremendous amounts of some fuel to produce this energy), but the burning of that fuel also causes pollution of one form or another. In addition to the air and/or water pollution generated at the generating facility, there are the wastes and other environmental effects associated with the mining or drilling for the fuels. These problems are often similar to those encountered in other mining operations.

Pollution Problems Due to Fuel Combustion
Most power currently is generated by the burning of coal (Figure 8-19). The typical environmental effects associated with a coal-fired plant are illustrated in Figure 8-20. A decreasing number of power plants are fueled by oil or natural gas. The typical emissions associated with these plants are illustrated in Figure 8-21 and 8-22. Figure 8-23 compares the major air pollutants according to fuel type.

FIGURE 8-18.   Interior view of coal-fired steam electric generator. (*Photo courtesy DOE.*)

FIGURE 8-19. Coal to be used as fuel for power plants. (*Photo courtesy Wisconsin DNR.*)

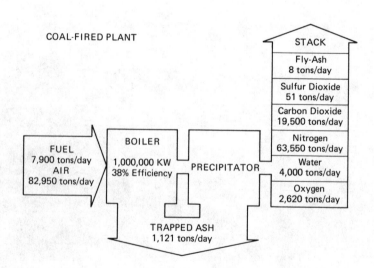

FIGURE 8-20. Environmental effects of coal-fired plant. (*Courtesy DOE.*)

184

GAS-FIRED PLAN

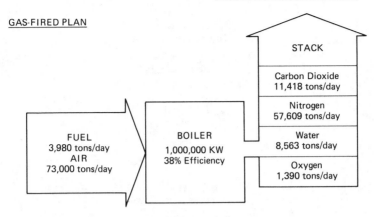

FIGURE 8-21.   Environmental effects of gas-fired plant. (*Courtesy DOE.*)

Nitrogen oxides are formed within most combustion devices, both by thermal fixation of atmospheric nitrogen and by oxidation of the nitrogen-containing organic chemicals within the fuel. The majority of the oxides are emitted as NO, which slowly converts to NO_2, a primary component of photochemical smog.

The most popular methods for NO emission control are 1) burning the fuel with insufficient air, followed by adding more air to complete the combustion, and 2) diluting the combustion air with recirculated flue gas, hence decreasing the peak flame temperature. Both techniques are most applicable to only those oxides produced by thermal fixation.

OIL-FIRED PLANT

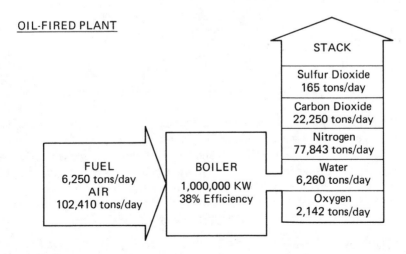

FIGURE 8-22.   Environmental effects of oil-fired plant. (*Courtesy DOE.*)

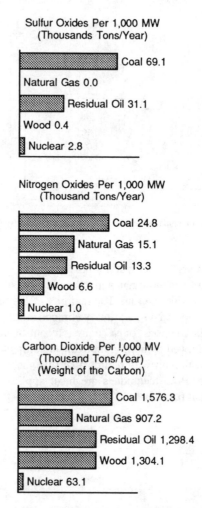

FIGURE 8-23.   Major power plant air pollutants, according to fuel type.

Exxon recently developed a process to limit the fuel oxides. NH_3 is injected into the furnace at a location such that the temperature is appropriate for a reaction to occur between the NO in the flue gas and the NH_3. A rapid gaseous reaction can occur, producing N_2 and water: $6NO + 4NH_3 \rightarrow 5N_2 + 6H_2O$. The temperature is critical, since the reduction of NO by NH_3 is very temperature-sensitive.[12]

Flue gas desulfurization, or the removal of SO_2 from the stack gas, can be accomplished by one of two general types of systems:[13]

1. "Throwaway" processes
2. "Marketable" processes.

The terminology refers to the residues of the cleanup processes. Throwaway systems produce a waste stream which must be disposed of in some manner; marketable systems produce a valuable product, such as sulfuric acid, elemental sulfur, or ammonium sulfate.

Throwaway systems are the more common, especially for large utility boilers. These techniques include both wet scrubbing and dry scrubbing methods.

Most systems today involve the use of wet stack gas scrubbers. About 90% of those scrubbers use some type of calcium absorbents, typically calcium hydroxide/calcium carbonate slurries, but these systems do exhibit difficulties due to the calcium sulfate precipitate that forms, plugging lines and fittings and forming a sludge that is difficult to dispose of.

A new technology using a proprietary, thermally regenerable organic amine salt as a sorbent was ommercialized by Union Carbide in 1991.[14] Their "CanSolv" process reportedly removes 99% of the SO_2 on a continuous basis, whereas limestone systems conventionally remove only 70–90%.

The basis of the CanSolv process is a scrubber, in which an absorbent is sprayed into the gas stream. The absorbent reacts with any SO_2, and then flows to a regeneration tower, where it is heated to liberate the SO_2. The SO_2 can then be dried and processed into sulfuric acid or elemental sulfur.

The major advantage of the process is its reversibility. Inorganic lime or limestone systems produce large quantities of waste sludge that require landfill disposal; the CanSolv process, with its amine salt, produces virtually no waste that must be disposed.

The first major demonstration plant was built in conjunction with Aluminum Co. of America (Alcoa), at its Warwick power plant near Newburgh, Indiana. Operating as an in-duct scrubber system, it removes SO_2 from half of the flue gas from an existing 150-MW boiler that burns 3.4% sulfur coal mined locally.

Other possible modifications of the wet scrubbers are outlined below.

1. An aqueous mixture of slaked lime, $Ca(OH)_2$, with added HCl forms a clear alkaline calcium chloride solution, $CaCl_2$. The SO_2 incident on this solution first generates calcium bisulfate, $Ca(HSO_3)_2$, which then oxidizes to gypsum, $CaSO_4$. The use of a clear solution rather than a slurry minimizes the plugging, though the acid does promote corrosion.
2. An NH_3 scrubbing solution can possibly be used, though the formation of ammonia salts has led to problems in that they are carried over as a vapor, forming a "blue plume."
3. A mixed solution of sodium citrate, citric acid, and sodium thiosulfate will also absorb SO_2. Contacting this solution with H_2S then releases elemental sulfur, and regenerates the sodium citrate.
4. Various sodium, potassium thiosulfate, and magnesium oxide absorbents are being tried, but the advantages of these are still questionable.

A number of other, dry processes are also currently feasible for SO_2 stack gas removal. Limestone ($CaCO_3$) and dolomite ($MgCO_3$), both plentiful and cheap, can be used to absorb the sulfur oxides. At the high temperatures encountered, the thermodynamics are favorable. For this process, pulverized $CaCO_3$ or $MgCO_3$ are injected into a boiler system. The heat drives off the CO_2 ($CaCO_3 \xrightarrow{heat} CaO + CO_2$), forming fairly reactive oxides. These oxides, CaO or MgO, can react with the sulfur oxides to form solid sulfites and sulfates, which are then removed by an electrostatic precipitator or some type of scrubber. A typical setup is shown in Figure 8-24. If a dry collection system (for example, a precipitator) is used, dust disposal is a problem; if a wet scrubber is employed, a saturated sludge with various soluble and insoluble sulfates and sulfites is formed, and a potential water pollution problem exists.

There are some other inherent difficulties in this method: typically, only about one-third of the stone reacts; also, stone from different parts of the country or different quarries react differently.

Catalytic conversion techniques can be used to eliminate SO_2. The SO_2 can be catalytically oxidized to SO_3 ($SO_2 + \frac{1}{2} O_2 \rightarrow SO_3$) and if this SO_3 is then allowed to react with water ($SO_3 + H_2O \rightarrow H_2SO_4$), the resulting H_2SO_4 mist can be removed and recovered by a mist remover. The H_2SO_4 is a useful byproduct but it does create some corrosion problems in the system.

Sulfur oxides can also be absorbed on the surfaces of $\frac{1}{16}$-in spheres composed of Al_2O_3 and sodium oxide (Na_2O), which are suspended in a fluidized-bed system. In a second unit, the sulfur oxides are removed from the spheres sphere by reacting the spheres with gas containing H_2 and CO, forming CO_2 and H_2S. The alkalyzed aluminum spheres are recycled, and the H_2S is converted to elemental sulfur. An advantage of this system is that the elemental sulfur byproduct can be stored outside until a market is available. A serious disadvantage, however, is the attrition of the

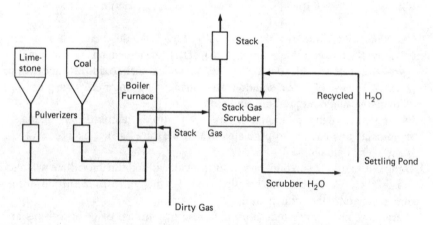

FIGURE 8-24.   Typical process for SO_2 removal by injection of pulverized limestone or dolomite.

spheres. Less expensive substitutes are being investigated, including oil shale, zinc calcine, and phosphate rock.

Various forms of chemical sorption are also under study—processes in which a liquid absorbs SO_2 from the stack gas, which is later regenerated as pure, dry SO_2. Though not yet commercially proven, three dry sorption systems are presently under development:

1. Activated carbon can be used to sorb the SO_2, which is then oxidized, in the sorbent, to H_2SO_4. The H_2SO_4-containing sorbent is transferred to another vessel, where the SO_2 is regenerated and then converted to elemental sulfur; the sorbent is recycled.
2. Fixed beds of cupric oxide (CuO)-coated alumina are also being tested. CuO can react with SO_2 to form cupric sulfate ($CuSO_4$). The $CuSO_4$ can then be reduced to elemental sulfur and regenerated CuO. Though the large-scale feasibility of this process is not yet demonstrated, it appears that it can operate at much lower temperatures than could the activated carbon processes.
3. A naturally occurring sodium bicarbonate mineral called nahcolite has also been tested as a possible absorbent. It can either be injected into the waste gas stream, or used to coat the inside of the fabric in the baghouse. The results have been mixed, usually requiring either a large excess of nahcolite or the recycling of a large amount of the gas. Disposal of the spent nahcolite might also lead to difficulties.[15]

The other major source of electrical power in the United States is nuclear power (Fig. 8-25). Currently, about 20% of the U.S. power is supplied by nuclear plants. In some localities, those that have no natural supplies of fossil fuels, nuclear power supplies a much greater percentage of the total power. For example, the Philadelphia Electric Co. generates 73% of its electricity from nuclear plants. Though a normally operating nuclear power plant produces no more radioactive emissions than does a large scale (1000 MWe) coal-fired plant, large quantities of highly radioactive waste fuel rods are produced that must be disposed of (Fig. 8-26). Proper long-term methods for disposal have not been agreed upon to date.

All of the fossil fuel generating plants, plus the nuclear plants, also create thermal pollution. Typically, 60–70% of the energy originally present in the fuel forms thermal pollution at the plant site. This added heat has the potential for seriously disrupting various ecosystems, particularly aquatic ecosystems.

Environmental Problems due to Extraction

In addition to the pollutants formed during the actual combustion of fossil fuels, and the radiation hazards of nuclear plants, very serious environmental effects are

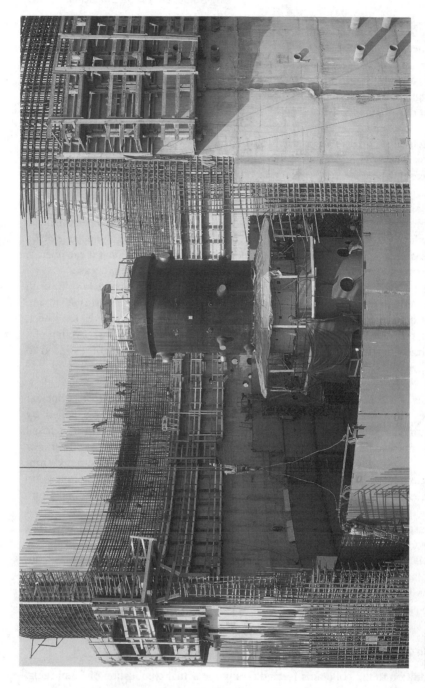

FIGURE 8-25. Construction of a nuclear power plant. Essentially all nuclear plants are distinguished by this dome shaped tower enclosing the actual reaction vessel (center). This architectural design is the strongest to withstand explosions from within and without, including air crashes and terrorists. *(Photo Courtesy of the Tennessee Valley Authority.)*

FIGURE 8-26.   Closeup of the reactor section of the Kewaunee Nuclear Power Plant. The fuel rods, approximately ¼″ in diameter and made of enriched uranium oxide, are placed among the boron control rods and immersed in water to produce the desired controlled thermonuclear reaction. The spent fuel rods are radioactive, and though the technology is developed for their safe disposal, this potential disposal is a political nightmare.

also produced during the fuel extraction procedures. Coal mining has, perhaps, the greatest environmental impact. Uranium mining creates some similar problems. Due to the much higher energy content per ton of fuel, the uranium mines are fewer and the environmental problems proportionately less. Drilling for petroleum and natural gas does have deleterious environmental effects, particularly offshore drilling. (The difficulties associated with oil production are discussed in Chapter

16, along with organic chemicals production.) Few power plants today are fueled by oil, due to its high cost.

The environmental effects of coal mining can be divided into five categories:

1. Land surface damage
2. Mine drainage
3. Refuse banks
4. Subsidence
5. Fires.

Land Surface Damage

Estimates are that, by the year 2000, about 240,000 km² (95,000 square miles—an area about the size of Oregon) will have been directly disturbed by mining. About half of the coal mined in the United States today is strip mined. Strip mining (Figs. 8-27, 8-28, and 8-29) consists of first removing the overburden from a narrow rectangular area, and placing it along (for example) the left side. The coal is then removed from that section. Next, the mining moves to a new area, immediately to the right. The overburden from the new area is dumped into the hole created by mining the old. This process is repeated, always moving in the same direction.

FIGURE 8-27.   Strip mining of peat, a very low grade coal, Connemara, Ireland.

FIGURE 8-28. Mountain strip mine near Oak Ridge, Tennessee. (*Photo courtesy DOE.*)

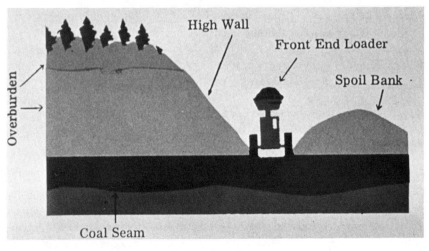

FIGURE 8-29. Simplified view of strip mine site. (*Courtesy DOE.*)

The result is a series of ridges and valleys. The major problem is that the topsoil usually ends up at the bottom, and the subsurface overburden on the top generally cannot support vegetation. If the strip mining occurs along the side of a hill, severe erosion, landslides, and siltation of nearby streams are likely.

Reclamation of these areas includes grading, fertilizing, and seeding. However, the overburden that was originally located near the coal deposits is invariably acidic and will not support plant growth. It must be well covered by other soil. The reclamation costs $1000–10,000/acre, approximately 5–10% of the value of the coal.

Acid Mine Drainage

Pyrite (FeS_2) is an impurity always present in coal, at least to some extent. This pyrite comprises a portion, but not all, of the sulfur content of coal. When FeS_2 makes contact with water, it reacts to form dilute H_2SO_4. This acid mine water is not serious for strip mines, for the pyritic materials can be isolated in the reclamation. For deep shaft mines, it is more difficult. The abandoned coal mines must be treated identically to abandoned iron mines: the water should not be allowed to enter; if it does enter, it should be moved rapidly. Treatment of the acidic water is possible at a cost of about $1/1000 gallons.

Refuse Banks

In deep shaft mining, (Figure 8-30) numerous waste materials are brought to the surface. Coal-cleaning creates more solid wastes (and some water pollution). Additionally, the burning of coal produces fly ash—typically, 15% of the coal by weight. These procedures have, in the past, created tons of waste materials. Current coal production generates 1.7×10^8 tons annually, enough to cover 1000 acres to a depth of 100 ft.

Not only are the refuse piles aesthetically unsightly, they also are subject to erosion by wind, water, and leach minerals, and they cover land that could be better utilized. The coal wastes are pyritic and thus acidic, and the fly ash is alkaline, so both can have major effects on nearby natural water systems.

The use of these materials to backfill abandoned deep shaft mines is a good idea, but generally it is too expensive. They can be used to some extent for construction purposes—roadbeds, bricks, etc. Improved coal-burning techniques will, it is hoped, decrease their production.

Subsidence

When the coal is removed by underground mining procedures, there is the potential for the collapse of the overburden, or subsidence. For active mines, this is minimized by leaving behind pillars of coal for support. Cave-ins are common, however, for abandoned mines, particularly in Pennsylvania. In urban areas, the streets cave and buildings are destroyed; in rural areas, the water flow

FIGURE 8-30.   Deep shaft mining of coal. A battery-powered tractor and trailer, articulated and very maneuverable, carry coal from mining machine or loader to conveyor belt or rail loading point. (*Photo courtesy of the National Coal Association.*)

patterns are upset, often causing water to drain into the mine, forming more acid mine water.

There are six million acres in the United States that have been undermined; one-third have subsided. The only way to prevent subsidence is by backfilling the mine, but, as mentioned, this is generally too expensive.

Fires

Fires occur in a substantial number of abandoned mines and refuse banks annually. Many are difficult to extinguish. These fires cause air pollution, destroy part of our coal resources, and contribute to the subsidence of underground mines by destroying the coal support pillars.

Hydroelectric Facilities

Hydroelectric generating facilities (Figure 8-31) are another possible power generating technique and are used fairly extensively in the aluminum industry. They are quite limited as to possible location—near a major river—and thus can never provide, nationwide, more than a small percentage of our total energy needs. To

FIGURE 8-31.   Norris Dam in Tennessee. This was the first of the TVA dams. (*Photo courtesy DOE.*)

generate power hydroelectrically, it is usually necessary to flood a large area for a reservoir, and frequently this is environmentally undesirable.

The hydroelectric facilities generate only minimal air or water pollutants, but the dams are obstacles to migrating fish. They change the flow rate of streams and the lake levels, both of which affect the aquatic ecosystems, and many organisms are destroyed by entrainment in the water flow into the turbine.

Conservation Techniques

Reduced Energy Consumption

Some aluminum plants are investigating techniques for reducing the energy consumption required for electrolysis. One likely technique being studied is to place a semi-conductor between the electrodes. Alcoa, on the other hand, has been developing a two-part smelting process. Alumina and chlorine are first converted to aluminum chloride, and the aluminum chloride is then electrolyzed to aluminum and chlorine. Not only does this process use 30% less electricity, but because it uses no fluorine, there is no need for collection and treatment of gaseous fluorinated byproducts. The chlorine is recycled directly to the first step, the conversion of the alumina to aluminum chloride. None of the techniques is yet at the commercial stage. But because of the large energy consumption, a decrease of only a small percentage of the total will have a major impact.

Aluminum Reclamation.[16]

Reclamation of aluminum is particularly important in terms of the energy conservation potential. The recycling of aluminum only requires approximately 5% of the energy needed to produce aluminum from the ore.

Recycled aluminum comes from three primary sources: aluminum scrap, foundry returns, and aluminum pigs. The scrap is frequently alloyed with other metals, so it must be melted and fluxed before reuse. Oils, grease, and paint are frequently burned off in a "chip dryer" prior to melting.

The scrap aluminum is usually melted in a reverberatory furnace equipped with a special charging well and a submerged hood (Figure 8-32). This well permits chips and other light scrap to be added to the melt from below the surface to prevent further oxidation. Fluxing gases such as chlorine gas, which is used to purge the metal of dissolved gases and magnesium, can be added in this well area. These gaseous fluxes can be collected and exhausted by the submerged hood.

Other "cover" fluxes are used to remove impurities and to cover the molten surface to prevent oxidation. These fluxes are usually salts, such as sodium or calcium chloride, or possibly cryolite, containing aluminum chlorides and fluorides. The fluxes cause the oxides and dirt to rise to the top of the molten metal, where they are skimmed off as "dross."

Air Pollutants

The major pollutants formed in aluminum reclamation are air pollutants. The chip dryer should remove all oils, greases, or paints, so smoke and odors should not be a problem of the reverberatory furnace. However, sodium, magnesium, potassium, and aluminum chlorides and fluorides are emitted from both types of aluminum fluxes. Aluminum and magnesium chloride are particularly troublesome, for they can sublime to a vapor, condense, and then react with water to form HCl and

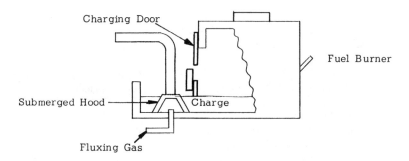

FIGURE 8-32.   Reverberatory furnace with charging well and submerged hood.

aluminum or magnesium oxide particulates. For every pound of chlorine used, about 15% is emitted as H_2, 30% as HCl, and the remaining 55% forms salts that are skimmed off the top of the metal.

The halogen gases used as fluxes are collected in caustic packed tower scrubbers. Caustic venturi scrubbers are another possibility, and would remove the particulates as well as the halogen gases.

Baghouses coated with a powdered alkaline adsorbent that can be used to adsorb and neutralize acid gases and simultaneously collect fine particulates are another possibility. About 800 lb of fine powdered adsorbent must be applied once a week to keep the system functioning properly.

There are thus a number of new techniques that show potential for energy savings in the aluminum industry. Not only would use of these methods save energy, however; most of them would also lead to an overall improvement in our environment.

References

1. Moore, Johnnie N. and Samuel N. Luoma, "Hazardous Wastes from Large-Scale Metal Extraction: A Case Study." *E S & T*, **24**, *No. 9*, September 1990, pp. 1278-85.
2. "Exploration for Copper in Wisconsin." *Skillings' Mining Review*, September 8, 1973, p. 7.
3. Black, William T., "Copper: A Colorful History and Bright Future." *ASTM Standardization News*, **15**, *No. 11*, November 1987, pp. 46-510.
4. Gardner, S. A. and Warwick, G. C. I., "Pollution-Free Metallurgy; Copper via Solvent Extraction." *McGraw-Hill's 1972 Report on Business and the Environment.* New York: McGraw-Hill, 1972, pp. 11-12.
5. Dasher, John and Power, Kenneth, "Copper Solvent Extraction Process: From Pilot Study to Full Scale Plant." *McGraw-Hill's 1972 Report on Business and the Environment.* New York: McGraw-Hill, 1972, pp. 11-15.
6. "SO_2 from Smelters: Three Processes Form an Overview of Recovery Costs." *Environmental Science and Technology* **4**, *No. 7*, July 1970, p. 554.
7. The South: A Resource Greater Than Oil." *Science News* **114**, *No. 10*, September 2, 1978, p. 116.
8. Lauber, Jack D., "Air Pollution Control of Aluminum and Copper Recycling Processes." *Pollution Engineering*, December 1973, p. 23.
9. "Monitoring Fluorides at an Aluminum Smelter." *Environmental Science and Technology* **11**, *No. 6*, June 1977, p. 550.
10. Cochran, Norman C., "Recovery of Hydrogen Fluoride Fumes on Alumina in Aluminum Smelting." *Environmental Science and Technology* **8**, *No. 1*, January 1974, p. 63.
11. "Red Mud Converted to Steel." *Chemistry and Engineering News*, February 26, 1979, p. 26.
12. "A Way to Lower NO in Utility Boilers." *Environmental Science and Technology* **11**, *No 3*, March 1977, p. 226.
13. McInnes, Robert and Ross Van Royen, "Desulfurizing Fluegases." *Chemical Engineering*, **97**, *No. 9*, September 1990, pp. 124-127.

14. Krieger, J., "Flue Gas SO_2 Removed by Organic Amine Salt." *Chemical and Engineering News, 69, No. 46*, November 18, 1991, pp. 7-8.

15. "New Scrubbers Tackle SO_2 Emissions Problems." *Chemical and Engineering News,* November 6, 1978, p. 24.

16. Lauber, p. 23.

9

Metals Finishing

Many metals, particularly iron and steel, undergo some "finishing" before commercial use. This finishing often consists of a coating of some other, corrosion-resistant metal such as zinc, copper, chrome, or nickel. Or, instead, the metal can be coated with some organic resin. Both types of coatings are very important, but they differ significantly in their use, application, and environmental problems, and thus should be considered separately.

METAL COATINGS

Application of the Coatings

Metal coatings can be applied in several ways. Considering, as an example, galvanizing, or the applying of a layer of zinc over iron or steel, after thorough cleaning, the iron or steel piece can be:

1. Dipped into a bath of molten zinc;
2. Coated with zinc dust and heated; or
3. Electroplated, typically using a cyanide bath to suspend the zinc ions.

Each of these processes produces a three-layer coating, the outer being pure zinc, the next the alloy $FeZn_7$, and the innermost layer $FeZn_3$. The two alloy layers are the thickest part of the coating. Other metals can be applied using similar techniques.

The most common of these finishing methods, not only for galvanizing but for the application of all metal coatings, is electroplating. In a typical electroplating process, the pieces to be plated are carried on conveyor racks, which are then dipped into various solutions. After the desired amount of plating has occurred, the racks are dipped into a series of rinse tanks. The water in these tanks flows countercurrent to the progress of the plated piece; hence, the rinse water in each of the successive rinse tanks is cleaner (that is, the piece is dipped into cleaner water with each rinse). The number of tanks varies with the process; three, for example, is common for chrome plating.

The water that is in the first rinse tank is by far the most contaminated. For a chrome plating, for example, up to 90% of the chrome plating metals originally in the plating solution may end up in the rinse water in this tank. This is the water that forms a good share of the effluent.[1]

Within the plating solution, various electrochemical reactions occur. For example, when plating chromium (Cr), the normal process starts with a bath containing Cr^{6+} ions. During plating, the Cr^{6+} is reduced to Cr^{3+}, then to metallic Cr. During the reduction, a film of these ions and the metallic Cr forms around the cathode, the object being plated. The Cr^{3+} ions readily form a stable six coordinated complex ion with water, but because of being bound in the film, they cannot react with water, and can instead be further reduced and deposited as Cr metal.

Typically, the plating solution also contains bisulfate ion, which functions as a catalyst for this procedure.

General Pollution Problems

The electroplating wastes are generated from all three phases of the electroplating process: 1) pretreatment, 2) plating, and 3) post-treatment.

Pretreatment

Pretreatment involves degreasing with soaps, alkaline cleaning, acid dipping, and/or "desmutting" to remove fine particles of base metal, particularly aluminum (Fig. 9–1).

Alkaline cleaners generally contain sodium compounds such as sodium hydroxide, sodium phosphate, sodium tetraphosphate, sodium carbonate, sodium silicate, and sodium metasilicate, and also a wetting agent such as kerylbenzene sulfonate. Their prolonged use for aluminum cleaning may lead to the formation of aluminum oxide precipitates. To increase the quality of the precipitate, citrates, tartrates, or gluconates may be added.

Smut, fine particles of base metal that adhere to the object being treated, is a problem with aluminum alloys of copper, silicone, or manganese. The desmutting agents typically contain sulfuric acid, nitric acid, phosphoric acid, chromic acid, or hydrofluoric acid.

Acid dips are somewhat similar, but instead generally contain sulfuric acid, nitric acid, phosphoric acid, chromic acid, hydrochloric acid, and/or fluoroboric acid.

FIGURE 9-1.   Pretreatment with soap to degrease a steel barbeque grill.

The wastewater from the pretreatment stage thus contains dissolved metals plus some of these cleaning agents.

Plating

The constituents of the wastewater from the plating operation will depend, of course, on the metals being plated and on the nature of the plating bath. Metals that are commonly plated include copper, nickel, chromium, zinc, cadmium, lead, tin, gold, and iron. Each of these can be plated from several types of baths. Table 9-1 lists the metals being plated, their usual purpose, and the type of plating baths used. Figure 9-2 illustrates a nickel plating bath.

Post-treatment

The post-treatment steps include processes such as chromating, phosphating, and coloring (Fig. 9-3).

Chromate coatings are applied to a number of metals for protection. The solutions contain chromic acid or its sodium or potassium salts, nitric acid, and some other organic or inorganic acids that function as activators or catalysts.

Phosphate coatings provide a good base for paints and other organic coatings. They also help resist corrosion or provide a base for rust-preventing waxes and oils.

TABLE 9-1 Plating Metals and Composition of Typical Plating Solutions

Metal	Purpose	Plating Solutions	Composition of Plating Solutions
Copper	Printed circuit boards, undercoat in decorative finishes.	Plain cyanide Rochelle cyanide Copper sulfate Copper fluoroborate	$CuCN$, $NaCN$, Na_2CO_3 $CuCN$, $NaCN$, Na_2CO_3, $NaOH$, Rochelle salt $CuSO_4 \cdot 5H_2O$, H_2SO_4 $Cu(BF_4)_3$
Nickel	Bright coating under thin Cr electroplate for decorative, corrosion- and wear-resistance purposes.	Watts Sulfamate Fluoroborate Chloride	$NiSO_4$, $NiCl_2$, Ni, H_3SO_3 $NiCl_2$, $(NiSO_3 NH_2)_2 Ni$, H_3BO_3 $NiCl_2$, $Ni(BF_4)_2$, Ni, H_3BO_3 $NiCl_2$, Ni, H_3BO_3
Chromium	Decorative or industrial finishes.	Chromic acid	H_2CrO_4, H_2SO_4 or H_2SO_4, F^-
Zinc	Protected iron and steel against corrosion	Cyanide Noncyanide baths	$Zn(CN)_2$, $NaOH$, $NaCN$, Na_2S_5 or Na_2S_4 $Zn_2P_2O_7$, $Na_4P_2O_7$, Na citrate, EDTA
Cadmium	Corrosion protection.	Cyanide Fluoroborate	CdO, Cd, $NaCN$, $NaOH$, Na_2CO_3 $CdBF_4$, Cd, NH_4CN, H_3BO_3
Lead or lead-tin alloys	Improve solderability, coating properties and performance of steels, Al, Cu, Cu-alloys.	Fluoroborate	Pb, HBF_4, H_3BO_3, glue, resorcinal, gelatin, hydroquinone
Tin or tin alloys	Improve solderability, corrosion protection, antifriction properties.	Sulfate Fluoroborate Halide	$SnSO_4$, H_2SO_4, gelatin, B-naphthol Gelatin, B-naphthol, Sn, HBF_4, H_3BO_3 $SnCl_2$ or SnF_2
Gold	Engineering (switches, semi-conductors), decorative.	Cyanide Acid	$AuCN$, KCN, K_2CO_3, K_2HPO_4, alloy metals $Au(CN)_3$, CN^-, citrates
Iron	Rare; for electroformed parts, dies, and cylinder liners.	Chloride Sulfate/chloride Fluoroborate	$FeCl_2$, $CaCl_2$ $FeCl_2$, NH_4Cl, $FeSO_4$ $Fe(BF_4)_2$, $NaCl$, H_3BO_3

[a]Data from Jacobsen, Kurt, and Laska, Richard, "Advanced Treatment Methods for Electroplating Wastes." *Pollution Engineering*, October, 1977, p. 43.

FIGURE 9-2.   A nickel plating operation takes place in a U-shaped tank in the foreground. Because the items to be plated are transported by an overhead conveyor system that can raise or lower them at the appropriate time, the length of the tank determines the plating time.

Generally, these solutions contain metal phosphates dissolved in phosphoric acid, with added nitrites, nitrates, chlorates, or peroxides as catalysts.

Coloring is caused by conversion of the metal surface to an oxide or other compound. Often, the colored surfaces are lacquered afterward. The pollutants can be any of a wide variety, but frequently include arsenic, antimony, mercury, molybdate, and permanganate.

Waste Handling

The majority of electroplating operations use cyanide baths of various formulations to suspend the metal ions. Thus, cyanide is a major waste of the metals coatings industry, one that is quite difficult to handle. Cyanides are versatile reagents, but they are, unfortunately, also toxic to bacteria in activated sludge, and so cannot be disposed of by ordinary treatment methods. The cyanide plating solution is "dragged over," to contaminate the rinse baths, as well as the spent plating solution itself. Other major contaminants present in the liquor and/or the rinse water from either the electroplating or other application processes are the various metal ions, zinc, cadmium, copper, nickel, hexavalent and trivalent chromium, various oils

FIGURE 9-3.   After nickel plating, these barbeque grills are given a chrome coating.

(from the metals cleaning), and possibly phosphates, depending upon the finishing process. Another difficulty may be large fluctuations in the pH. There are a number of general ways the electroplating wastes can be handled. The oils are usually skimmed from the surface and recovered. They could also be combusted. The phosphates can be removed by conventional methods, as they are in municipal treatment plants.

The "climbing film evaporator" is one new, reportedly flexible and efficient method that can be used to recover a variety of the process chemicals. The rinse water continuously flows into an evaporator. Low-pressure steam, fed into a vertical tube containing the rinse water, vaporizes that rinse water, and drives a film of the concentrated solution upward, into a vapor-liquid separator. Within the separator, the heavier materials (such as nickel and chrome) settle out; the water vapour continues upward, enters a filter, and then a condensor. This water can be again used in the rinsing process. The concentrated contaminants are drained into a storage tank, and can later be reused in the process (for example, in the plating tank). Typical recovery efficiencies are about 80–85% for chrome, with smaller percentages for the chrome treatment chemicals.[2]

Though there are many metals finishing plants that dispose of their heavy metals by hazardous waste landfill or similar ways, a large percentage of these metals are recovered by some technique. There are other recovery processes available, often designed for specific chemicals.

Specific Chemical Recovery Systems

Hexavalent chrome is a particular problem because it is toxic. It is often produced by the anodizing of aluminum by dipping it into chromic acid (H_2CrO_4) and chromate salt solutions. When the aluminum is rinsed, the residual hexavalent chrome is washed from the surface.

There are several possible treatment methods. The waste stream containing the hexavalent chrome and some suspended solids and floating materials can be treated with antifoam and coagulating agents. The waste water then enters an aeration tank. After aeration, the suspended and floating materials are removed, and the sludge is allowed to settle. The hexavalent chrome is absorbed by the various coagulating agents and is thus predominantly removed from the effluent. Any residual can be reduced to nontoxic trivalent chrome by use of ferrous chloride,[3] sulfur dioxide, sodium bisulfite, or ferrous sulfate.[4] Or the chrome rinse waters can be passed through both cation and anion exchange resins. The trivalent chrome, together with the copper, nickel, and similar substances, are exchanged on the cation resin, releasing sodium ions. The hexavalent chrome, because it occurs as the oxide ions, CrO_4^{2-}, is exchanged on the anion resin. After saturation, the cation exchanger is regenerated with acid and the anion exchanger with sodium hydroxide. The regenerating effluent from the cation exchanger, containing some acid and the metal ions, is collected and neutralized; that from the anion exchanger, containing sodium chromate and some sodium hydroxide, can be converted to chromic acid by passing it through the cation-exchange resin. The net effect is to 1) regenerate the chromic acid, and 2) prepare the cation resin for further exchange of sodium-heavy metal ions.[5]

Chromic acid fumes are a serious problem in plating systems that involve chrome plating. This mist is emitted at 180°F, and has a size in the submicron range. The typical concentration, as emitted from the tanks, is 6–7 mg/m³, well above the accepted level of toxicity of 0.01 mg/m³.[6] This mist can be treated by a three-stage process consisting of wet scrubbers followed by electrostatic precipitators and then an activated carbon adsorption system.

Metals such as trivalent chromium, nickel, copper, and zinc can be, as discussed above, collected by ion exchange. Another possible treatment method involves pH adjustment to convert the soluble heavy metals to insoluble metal hydrates. These can then be removed from the effluent by sedimentation, clarification, and/or filtration.[7] Flocculating agents such as ferric chloride, aluminum sulfate, or a polyelectrolyte can be used to aid in the sedimentation. Reverse osmosis is yet another possibility. Various plating baths have been successfully treated by reverse osmosis techniques. These include Watts nickel, nickel sulfamate, copper pyrophosphate, nickel fluoroborate, zinc chloride, copper cyanide, zinc cyanide, and cadmium cyanide. On the other hand, chromic acid and very high-pH cyanide baths are not suitable for reverse osmosis treatment.[8]

Cyanide Decomposition[9,10]

The cyanides are also very toxic, and must generally be destroyed rather than recovered. There are several effective methods available for this.

The most frequently employed process for cyanide decomposition is alkaline chlorination. This can be accomplished by one of two ways. One method is by direct addition of sodium hypochlorite (NaOCl) to the waste water. The NaOCl oxidizes the cyanide to cyanate: $CN^- + OCl^- \longrightarrow OCl^- + Cl^-$. This process can, alternatively, be accomplished by the addition of chlorine gas and sodium hydroxide to the wastewater. The chlorine and sodium hydroxide react to produce NaOCl: $Cl_2 + 2NaOH \longrightarrow 2NaOCl + H_2$. The NaOCl then reacts as above.

The cyanates that result are less toxic than cyanides, though they are not totally desirable. Lowering the pH of the cyanate solution to 2 or 3 will hydrolyze the cyanates to CO_2 and NH_3: $H_2O + H^+ + OCN^- \longrightarrow CO_2 + NH_3$. This procedure, however, requires the subsequent neutralization of the exceedingly acidic solution by a base before discharge.

Chlorine gas can be used similarly. The chlorine can convert cyanide to cyanate:

$$Cl_2 + NaCN \longrightarrow CNCl + NaCl$$
$$CNCl + 2\ NaOH \longrightarrow NaCNO + H_2O + NaCl.$$

The reaction proceeds most rapidly at pH's above 9.0.

Potassium permanganate ($KMnO_4$) is a strong oxidizing agent that can also oxidize cyanides. The reaction occurs as: $2KMnO_4 + NaCN + 2KOH \longrightarrow 2K_2MnO_4 + NaCNO + H_2O$. It is necessary to maintain the pH between 12 and 14.

Wet oxidation in the aqueous phase at elevated pressures and temperatures, absorption of the cyanide by activated carbon, followed by oxidation with O_2 and a copper catalyst, and ozonation of the cyanides are other possible methods of decomposing the cyanides.

When cyanide concentrations are very high, electrolytic oxidation is an available treatment method. The effluent is electrolyzed for 1–2 days at temperatures around 200°F. The cyanides are oxidized to CO_2 and NH_3. Initially, the reaction is very rapid, but it becomes progressively slower with time, due to the formation of a conducting solution, and difficulties may be encountered.

The "Kastone process," developed by E. I. DuPont de Nemours and Company, is another cyanide treatment processes. Kastone is a proprietary chemical, consisting of an aqueous solution of about 41% hydrogen peroxide (H_2O_2), a small amount of stabilizers, and a small amount of a catalyst that is designed to break apart metal-cyanide complexes. The cyanide wastes are agitated and heated to about 120°F; then a 37% solution of formaldehyde (HCHO) and the Kastone are added. The result is a series of various very complex products, plus some metallic oxides, hydroxides, and carbonates, most of which form a floc and can be filtered. Examples of other compounds formed include cyanate ions ($CN^- + H_2O_2 \rightarrow CNO^- + H_2O$), glycolonitrile ($CN^- + HCHO + H_2O \longrightarrow HOCH_2CN + OH^-$),

formate ion and ammonia ($CH^- + 2H_2O \xrightarrow{H^2O^2} HCOO^- + NH_3$), and glycolic acid amide ($HOCH_2CN + 2H_2O_2 \longrightarrow HOCH_2COONH_2 + H_2O$).

There have been tests of additional cyanide detoxification methods, as well. These include ion-exchange methods and radiation detoxification by gamma rays. The latter can destroy the carbon-nitrogen triple bond, converting the cyanide to CO_2 and nontoxic nitrogenous products.[9,10]

Some attempts have been made for cyanide recovery. Vacuum evaporation has been proposed for cadmium cyanide solutions. Reverse osmosis is being considered for copper or zinc cyanide baths.

As the pollution control restrictions have increased, particularly those relating to hazardous substances, more thorough collecting and treatment of these wastes has been necessary; it is likely that more recovery will also be instituted, if, for no other reason, because of the economic advantages of doing so.

The IBM Federal Systems Division Facility at Owego, New York[11]

The IBM Federal Systems Division operation at Owego, New York has facilities for the plating, cleaning, surface treatment, and etching of various metals, especially copper. These operations generate approximately 500,000 gpd of wastewater. The wastewater generated at the main facility is separated from the beginning into three drain systems: 1) general rinse (all nonchrome, noncyanide rinse waters), 2) acid/alkali (small volume, intermittent acid and alkaline cleaning solutions, nonchrome and noncyanide spent plating bath waters), and 3) chrome (chrome rinse waters plus, rarely, a concentrated chrome dump). A fourth drain outside the main facility handles the ion-exchange regeneration wastes (the demineralizer waste) and the backwash from the sand filters. The additional collection tanks handle waste ferric chloride, and waste copper baths and rinses, respectively. Further, the photographic wastes and the sanitary sewer wastes are discharged into the Owego sewage system, to be treated by the municipal treatment plant by biological oxidation procedures.

The General Rinse System
The bulk of the wastewater is treated by the general rinse system. This water is from all rinse operations except those containing chrome or cyanide. The wastewa-

ter in this drain system is first routed to an equalization basin, to reduce concentration and flow variations.

From the equalization basin, the water flows to a settling basin, where the oil is skimmed from the top. This basin can act also as a water-solvent separator should there be any chlorinated hydrocarbon spills.

Next, the wastewater flows to a second equalization tank for a limited additional equalization, and then to the demineralizer/neutralization tank. In this latter tank, the pH is raised from 6-7 to 11 by the addition of lime: ferric chloride is added as a coagulant and hydrogen sulfide gas is bubbled in to break up any complex copper ions that might be present due to spills.

From the demineralizer/neutralization tank, the water flows to two clarifiers. These clarifiers constitute the most important phase of the treatment, and remove the metallic hydroxides and suspended solids by means of a hydraulically suspended sludge "blanket." The blanket is formed within the clarifiers by combining a polymer substance with recirculated sludge (to seed the sludge buildup) and some of the influent from the previous process. The wastewater enters from below, passes through the sludge blanket, where the metallic hydroxides and suspended solids are filtered out, and then goes upward out of the clarifier to a post-neutralization tank.

In the post-neutralization tank, sulfuric acid is added to lower the pH to 6.5-8.5. The water then flows to a final lagoon, where it is mixed with water from the demineralizer operation, and allowed to undergo another equalization operation before discharge to the river.

The Acid/Alkali System

The acids and/or alkalies from the various baths and cleaning operations are treated in two collection tanks. While one tank is collecting the wastewater, the second is treating and discharging. Normally, the pH of this waste is low (acidic), so lime and/or sodium hydroxide are added to raise the pH to 9 or 10. From the tanks, the wastewater is shunted to the sludge lagoon, where the solids settle out. The supernatant is further treated in the general rinse system, and eventually discharged to the final lagoon and then to the river.

The Chrome System

The chrome-containing wastewater requires a separate treatment procedure to detoxify the Cr^{6+}. This wastewater is sent to one of two collection tanks, similar to those used in the acid/alkali system. The pH of the water is lowered to 2 or 3 by addition of sulfuric acid; sodium bisulfite then is used to reduce the Cr^{6+} to Cr^{3+}. Lime or sodium hydroxide returns the pH to 8.0-8.5. The water flows to the sludge lagoon to remove the solids, and the supernatant is combined with that of the acid/alkali system and treated by the demineralizer/neutralization tank and by the clarifier of the general rinse system, and then is discharged to the river.

The Demineralizer Waste System

The waste from regeneration of the ion-exchange system consists of sulfuric acid or sodium hydroxide wastes. This water volume is large, and there are many pH extremes. The water is first routed to a holding/equalization pond, where the pH is adjusted to 7.0 or 7.5 by the further addition of either sulfuric acid or sodium hydroxide, as appropriate. It then flows to the final lagoon, where it is mixed with the effluent from the clarifier, equalized, and discharged to the river.

The backwash from the sand filters is discharged directly into the final lagoon for settling of the suspended solids. In the future, it may be discharged to the sludge lagoon to avoid disturbing the solids in the final lagoon, and the supernatant treated as with the acid/alkali and chrome wastes.

Effluent Quality

This system has been in operation for a number of years, and has operated very successfully. The levels of copper, iron, chrome, total suspended solids, oil, grease, pH, and NH_3 are well within the current EPA guidelines.

CHEMICAL (ORGANIC RESIN) COATINGS

The Nature of Chemical Coatings

Chemical coatings are most familiar to the majority of people as "paints." Chemical coatings are, specifically, those paints that are used for industrial applications. Typical applications include automobiles, appliances, house sidings, and food and beverage cans.

Almost all chemical coatings are composed of a solid coloring matter suspended in a liquid. All require a resinous binder that can be converted to a solid film upon disposition. Most contain a solvent to lower the viscosity of the binder. This solvent evaporates after disposition of the film. Most coatings also contain pigments to impart opaqueness or color. These pigments can be either organic or inorganic compounds. Chemical coatings may also contain defoamers, thickeners, driers, and/or flow agents to improve specific characteristics.

The protective properties of a coating are mainly determined by the binder. Liquid coatings can be of two types: dispersions or solutions. Solutions occur when the binder is actually dissolved in the solvent. Dispersions occur when the binder forms tiny spheres ($10\mu m$ in diameter) that are suspended in the solvent. In both cases, when the solvent evaporates, the resin binder and any pigment that is present remains and fuses into a continuous film. Table 9-2 lists some common pigments, binders, and solvents.

TABLE 9-2 Typical Pigments, Binders, and Solvents[a]

Pigments	Binders	Solvents
Iron oxides (Fe_2O_3, Fe_3O_4)	Oils	Hydrocarbons
Lead chromate ($PbCrO_4$)	Alkyds	Mineral spirits
Lead molybdate ($PbMoO_4$)	Nitrate and acetate celluloses	Benzene
Zinc chromate $4(ZnO \cdot K_2O) \cdot 4(CrO_3) \cdot 3(H_2O)$	Acrylics	Toluene
Lead oxide (PbO)	Vinyls	Xylene
Cadmium sulfide	Phenolics	Alcohols
Cadmium sulfoselenides	Epoxies	Methanol
Cadmium selenides	Polyurethanes	Ethanol
Chromium oxides (Cr_2O_3, $Cr_2O_3 \cdot 2H_2O$)	Silicones	Butanol
	Amino resins	Ethers
	Styrene-butadiene	Dimethyl ether
	Polyvinylacetates	Ethylene glycol
		Ketones
		Acetone
		Methyl ethyl ketone
		Esters
		Ethyl acetate
		Butyl acetate
		Tetrachloroethane
		Nitromethane
		Water

[a]Data from *Riegel's Handbook of Industrial Chemistry*, 7th Edition, James A. Kent (Ed). New York: Van Nostrand Reinhold, 1974, pp. 655–659.

Application of Chemical Coatings

Chemical coatings can be applied by a variety of techniques. Electrostatic spraying (Figure 9-4) capitalizes on the use of electrostatic forces. The object to be coated must be conductive. An electrostatic potential is formed between the object and the atomized paint droplets, so that the droplets are attracted to the object, forming a uniform coating on even irregular shapes.

Dipping the object into the coating is suitable for undercoating where uniformity and appearance are not important. Flow coating, where the coating is simply allowed to flow over the object, is another possibility.

Environmental Problems[12]

The major environmental problem faced by any industry using chemical coatings is the emission of the solvents during application and curing. Most coatings currently in use are only 25-45% solids by volume. The remainder of the coating is primarily solvents. The organic solvents are highly undesirable as air effluents. All of the hydrocarbons have been determined by the EPA to be photochemically

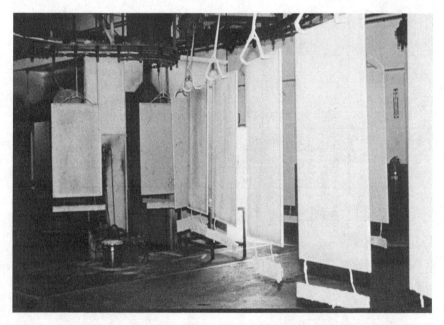

FIGURE 9-4.   Electrostatic spraying for the application of chemical coatings. (*Photo courtesy Lilly Industrial Coatings.*)

reactive, at least to some degree, and therefore are unacceptable for discharge to the atmosphere. Many solvents are also odorous.

The EPA has issued several sets of guidelines for specific industries using chemical coatings. These guidelines have been given to, for example, the automotive, can, coil, fabric, and paper coating industries.

Methods Used to Reduce Solvent Emissions

The methods currently being employed and/or investigated to allow the various industries to meet these guidelines fall into two general categories: 1) removal of the pollutants from the air stream before its discharge, and 2) modification of the coatings so that they contain fewer potential pollutants. The use of the first technique, the removal of the pollutants after their formation, is primarily the responsibility of the various industries that actually use the coatings. Modification of the coatings to generate fewer pollutants, on the other hand, is the responsibility of the chemical coatings industry itself. Both the user and the supplier must work together to meet the environmental regulations economically while still maintaining a high-quality product.

Effluent Gas Clean-up

There are two major techniques currently available for removal of the solvent emissions: incineration and activated carbon adsorption. Incineration of the solvents generally requires supplemental natural gas; therefore, the operating cost is very high. In some regions of the country, natural gas is periodically in short supply. If the solvents could be recovered and reused, there would be significant additional savings realized.

Carbon adsorption is the only method currently being used to any extent for solvent collection and recovery. Carbon adsorption of vapors is a technology that has been in existence for over 60 years.[13] Its use is not restricted to the industries using chemical coatings, but includes various chemical plants (making pharmaceuticals, synthetic rubber, etc.), rotogravure printing companies, and paper coating and similar industries.

Figure 9-5 illustrates a typical carbon adsorption system. The activated carbon, with an extremely large surface-to-volume ratio, is capable of adsorbing organic solvents on its surface. After a few hours of collection, the solvents must be removed from the pores of the activated carbon to 1) regenerate the carbon and 2) recover the solvents. Typically, two carbon beds are in the adsorption mode while a third is being regenerated, as illustrated.

There are three methods for desorption commonly used. The "classical" technique, used on about 99% of all carbon adsorption systems, is to run live steam through the carbon bed counterflow to the normal solvent flow direction. A hot inert gas, such as nitrogen, can also be circulated to remove the adsorbed molecules.

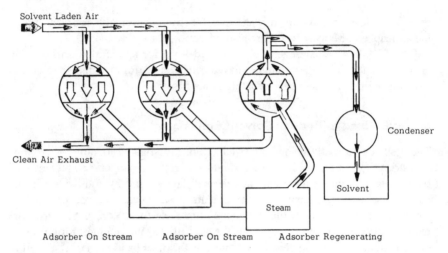

Solvent Laden Air

Condenser

Clean Air Exhaust

Solvent

Steam

Adsorber On Stream   Adsorber On Stream   Adsorber Regenerating

FIGURE 9-5.   Schematic of a typical carbon adsorption system.

Application of a vacuum to the carbon beds will also desorb the solvents, though the cost is high due to the capital investment required to allow the use of vacuum equipment on each and every adsorber.

Reactivation of the carbon beds is also required periodically, to restore the carbon to its original condition. Reactivation usually is done at a carbon manufacturing plant, not on site, and involves heating the carbon to about 1500°F in a reduced O_2 atmosphere. This procedure removes any contaminants from the carbon pores. The frequency of reactivation varies with the plant and the industry, and depends on the quantity of trace contaminants such as plastisers and high boiling point organics in the solvent stream, but reactivation would typically be necessary at periods of 3–15 years.

Corrosion is another problem often experienced with carbon adsorption systems. Many solvents are noncorrosive (toluene, xylene, heptane, hexane, and the alcohols), but many esters, ketones, and chlorinated solvents can develop acidic properties when exposed to sunlight and/or traces of water. This is particularly pronounced in the presence of activated carbon, whose large surface area can act as a catalyst. Though less than 1% of the solvent is converted to acid, the construction materials must be such so as not to undergo corrosion. Stainless steel is frequently recommended.

The screen that supports the carbon bed is generally about 6 mesh in size (openings of about 3.36 mm), and thus is comprised of very fine wire. A galvanic cell can develop between the screen and the carbon, leading to corrosion of the screen, even if the solvents present are not corrosive. Use of a high-quality stainless steel screen also eliminates much of this problem.

Modification of the Coatings[14]

Modification of the coatings so that they produce fewer solvent emissions is another tack that has been taken by the industry in recent years. These new coatings fall into several classes: waterborne coatings, high solids coatings, coatings that can be cured by ultraviolet (UV) light or electron beams, powder coatings, and the application of the coatings by electrodeposition.

Waterborne coatings: Waterborne coatings are now used in a large percentage of all industrial applications. Some examples include automotive primers and beer and beverage cans. Waterborne resins have been used for many years in household applications (latex paints), but only recently have they been used to any extent for industrial applications.

There are three types of waterborne resins: emulsions, dispersions, and solubles. The emulsions are aqueous dispersions of resin similar to that now used in house paints. Dispersions are colloids of ultra-fine particle size resins in water. The water-soluble or water-reducible polymers form true solutions. The emulsions, though the resins are of high molecular weight with reasonable viscosity and solids content, do not provide a high gloss, though their mechanical properties are outstanding. The solutions have limited storage stability since the resins tend to hydrolyze and decompose in six months to a year. They also tend to cause blister problems due to entrapment of water. All three types tend to be more water-sensitive, and may not have adequate salt spray or humidity resistance, though this is most pronounced for the dispersions and solutions. Another problem with the waterborne resins is the additional energy required for curing. In spite of these problems, more and more waterborne coatings are being utilized, and better characteristics are being realized, as the need to eliminate solvent emissions promotes increased research in this area.

High-solids coatings: Most traditional solvent-based coatings are 25–45% solids by volume. Though there is some controversy about what constitutes a "high-solids" coating, a regulation from California states it must contain at least 80% solids by volume (approximately 90% by weight). Some of the coatings today that meet current EPA standards are 70% by volume.

As the solids content of the coating is increased, the polymer size of the resin must decrease. This tends to detract from the good film properties. This forces careful polymer choice, or possibly using medium molecular weight polymers combined with heat to reduce the viscosity during application.

A high-solids coating does have cost advantages. The solvent plays no part in the final coating, so one usually pays for something that is later removed. Storage, shipping, and handling costs are also reduced due to decreased volume for the same usable material. Finishing costs are also lower: less solvent means less air consumption and lower fuel costs.

The disadvantages of high-solids coatings include problems of flow, cratering, wetting, and sagging (Figures 9-6 and 9-7). Layers of 1–2 mils are difficult to achieve, yet frequently desired. Gloss control is also difficult.

FIGURE 9-6. Cratering effects on high solids coatings. (*Photo courtesy Lilly Industrial Coatings.*)

FIGURE 9-7. Sagging of chemical coatings. (*Photo courtesy Lilly Industrial Coatings.*)

Some industries (extrusion finishings on aluminium, steel building components, selected automotive steel parts) have adopted high-solids coatings, though much research and development must yet be done on them.

Other Possibilities

Electrodeposition from a water bath containing the coating is a possibility. Only one thin layer of coating can, however, be applied by this method, and the expense is high. Most automobiles are primed using this technique.

Curing the coating by UV light or electron beams is under study, but currently this is not economically feasible. The capital equipment cost for the electron beam coatings is particularly high.

Rather than using a liquid, a powder may in some instances be employed. The dry powder can simply be spread on the subject and then fused. One disadvantage to this, however, is the inability to achieve thin films.

The direction any particular industry should take to meet the EPA solvent emission guidelines can thus vary, depending on its unique circumstances. In the use of chemical coatings, many options are available. There are numerous trade-offs between energy consumption, economy, and performance standards that must be considered.

References

1. Bhatia, Sulim and Jump, Robert, "Metal Recovery Makes Good Sense!" *Environmental Science and Technology* **11**, *No. 8*, August 1977, p. 752.
2. *Ibid.*
3. "Manufacturers Stress Benefit of Recycling After Treatment to Conserve Water and Yield Products." *Business and the Environment.* New York: McGraw-Hill, 1972, pp. 14–21.
4. Jacobsen, Kurt and Laske, Richard, "Advanced Treatment Methods for Electroplating Wastes." *Pollution Engineering*, October 1977, p. 43.
5. *Pollution Control Technology.* New York: Research and Education Association, 1973, p. 452.
6. Gothard, Nicholas, "Chromic Acid Mist Filtration." *Pollution Engineering*, August 1978, p. 36.
7. Zievers, James and Novotny, Charles, "Curtailing Pollution from Metal Finishing." *Environmental Science and Technology* **7**, *No. 3*, March 1973, p. 209
8. Jacobsen and Laske, p. 46.
9. "New Process Detoxifies Cyanide Wastes." *Environmental Science and Technology* **5**, *No. 6*, June 1971, p. 496.
10. Goldstein, Mel, "Economies of Treating Cyanide Wastes." *Pollution Engineering*, March 1976, p. 36.
11. Forbes, John M., Jr., "IBM Owego Gives Metal Finishing Wastes Total Treatment." *Pollution Engineering*, March 1979, p. 46.
12. Bailey, Robert S., Lilly Industrial Coating, Inc. "Technology Turmoil in Industrial Coating." Presented at a Chemist Coaters Meeting in Minneapolis, January 25, 1979.

13. Worral, Michael J., American Ceca Corporation. "Solvent Recovery II—Desorption and Corrosion." Presented at GRI/GTA Seminar on "Emission Control for Packaging and Specialty Gravure Printers," January 11, 1979.
14. Bailey.

10

Cement Manufacture

The manufacture of cement is a large industry in the United States—and in the world (Figure 10-1). It is also potentially a very dusty (ninth in the United States in potential air pollutants[1]) and very energy-consuming (sixth in the United States[2]) one. Many programs have been undertaken in recent years to modernize the processes and equipment, but, as was seen also in the steel industry, many

FIGURE 10-1. The Holderbank Management and Consulting cement plant located in Holderbank, Switzerland.

of the U.S. plants are fairly old, and the capital investment required for many of the process conversions is very high. The cement manufacturing process consists of taking raw materials such as limestone, shale, sand, clay, iron ore, and fly ash, grinding them finely and mixing them in proportions so that, after mild heating to drive off any water and CO_2 available in the limestone ($CaCO_3$), one obtains typically 64% calcium oxide (CaO), 22% silicon dioxide (SiO_2), 3.5% aluminum oxide (Al_2O_3), and 3.0% iron ore (as Fe_2O_3). These raw materials are processed at very high temperatures and react by solid-solid reactions to form four cement compounds: tricalcium silicate ($3CaO \cdot SiO_2$), dicalcium silicate ($2CaO \cdot SiO_2$), tricalcium aluminate ($3CaO \cdot Al_2O_3$), and tetracalcium aluminoferrite ($4CaO \cdot Al_2O_3 \cdot Fe_2O_3$). The exact proportions of these final products determine the cement characteristics; for example, the hardening time, the early strength, and the final strength.

The overall cement process is diagrammed in Figure 10-2. The largest volume raw material is $CaCO_3$ or comparable materials (such as oyster shells in locations where appropriate). The $CaCO_3$ as mined is often in chunks up to 30 inches in diameter. These must be crushed to about ⅜ inches, and then mixed with the sand, shale, and other ingredients, for grinding to about 60 μm in diameter. Frequently, the initial crushing is done at the quarry, prior to transport to the cement plant (Figures 10-3 and 10-4). After grinding, depending upon the exact process, water may be added. The mixture is then taken to some high-temperature processing unit for conversion to cement clinker. The clinker must be cooled before further processing. Then it can be either stored, sold, or shipped as is, in chunks ¼ inch to several inches in diameter, or it can be ground with gypsum and other possible additives to a fine powder, the finished cement. The cement can be bagged or sold in bulk to distributors.

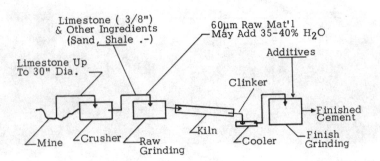

FIGURE 10-2.   The overall cement process.

FIGURE 10-3. Barging the raw materials from the quarry to near the Signal Mountain, Chattanooga, Tennessee cement plant.

FIGURE 10-4. Conveying the raw material from the barge to trucks for hauling to the plant.

CEMENT PROCESSES

The various types of cement manufacturing processes can be classed as wet, dry, or semi-wet. Wet processes currently account for about 60% of the U.S. cement capacity, with dry (and, to some extent, semi-wet) processes becoming continually more prevalent. Wet processes are particularly useful when the raw materials are wet as quarried. The addition of water also permits easier and better mixing of the raw materials. Dry processes consume significantly less energy, and often can handle any particulate emission problems more easily. However, because of the large capital investments, it is currently uneconomical for a wet process to be converted to dry for these reasons alone.[3]

Wet Processes

A typical wet rotary kiln is shown in Figures 10-5 and 10-6. The raw material, after being mixed with 30–40% water, is fed into one end of the kiln. As the kiln slowly rotates, the material migrates toward the hotter flame end, getting hotter and hotter,

FIGURE 10-5.   Wet process rotary cement kilns. Note the stacks at the end.

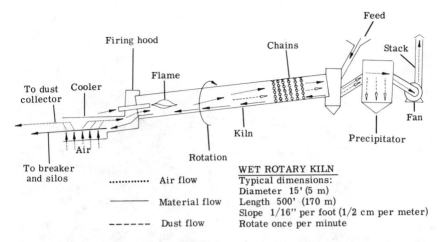

FIGURE 10-6.   Diagram of a typical wet process rotary kiln.

as illustrated in Figure 10-7. The typical rotary kiln uses pulverized coal as a fuel (Figures 10-8 to 10-10). (Prior to the 1973 energy embargo, many plants employed natural gas and a few used fuel oil.) Initially, the material becomes dehydrated due to the vaporization of the water. The next step is that of "calcination": the driving off of the CO_2 from the $CaCO_3$. The reaction proceeds as $CaCO_{3(s)} \rightarrow CaO_{(s)} + CO_{2(g)}$. By the time the materials reach the flame area, they are white hot (2600–2800°F), and a series of solid-solid reactions occurs by diffusion. These lead to "clinkerization," the formation of the cement clinker. After this extremely hot area, the temperature drops, and the clinker starts to cool. The materials finally then drop

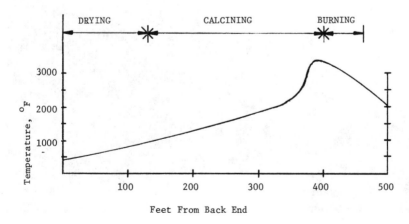

FIGURE 10-7.   Temperature profile within a kiln.

FIGURE 10-8. Coal to be used as kiln fuel.

FIGURE 10-9. Conveyor system for transport of coal to plant (Holderbank, Switzerland).

224

FIGURE 10-10.   Front end of wet process kiln. Note the large coal fuel pipe from the left that bends around, then enters the center of the kiln. The operator's room is located on the extreme right.

out of the kiln onto a cooler, through which large volumes of relatively cool air are passed.

The air from the cooler, rather than being "wasted," is channeled into the kiln for use by the flame. This air, in traversing the kiln, becomes turbulent, and often picks up some of the finer raw materials particles. These become entrained in the air stream. The air simultaneously transfers heat to the back end of the kiln.

Before the air can exit the kiln, it is passed through a dense curtain of chains. The chains serve two purposes: 1) they remove some of the entrained dust, and 2) they act as a mechanism for heat transfer, to retain as much as possible within the kiln.

The air, after leaving the kiln, is ducted to an electrostatic precipitator for particulate removal, and then ducted to the stack. The particulates are 500-600°F; hence, the use of fabric collectors has been often impractical, though a few plants do use them. Essentially all the dust particles are collected; the question then becomes one of disposal.

Dry Processes

A long kiln similar to that used in the wet process can also be used for a dry process. This is not very common today, however, for these long dry kilns are not very efficient. It is estimated that more than twice the dust is entrained in dry process kilns as in comparable wet process kilns.[4] Moreover, though a dry kiln consumes less energy per ton of clinker than does a wet process, other dry processing techniques are far superior.

The preferable dry processing method is by a suspension preheater system (Figure 10-11). The finely ground dry raw materials are fed into the preheater at the top, countercurrent to the air flow. This air flow originates in the cooler and thus has been heated by traversing that cooler and also a short rotary kiln section before being ducted to the preheater. Hence, it is sufficiently hot to not only preheat, but also to partially calcine the incoming materials. The physical arrangement in the preheater of a series of cyclones is such that the hot air and the feed can have intimate contact in a series of stages for maximum heat transfer and optimum efficiency. The addition of a "flash calciner," a stationary furnace interposed between the rotary kiln and the suspension preheater, increases the amount of calcination that occurs within the preheater, thus increasing the potential capacity of the rotary kiln.

The hot particulate feed, after passing out of the preheater and the flash calciner, enters a short rotary kiln where it undergoes clinkerization. The need to accomplish only this last stage of the processing within the kiln is more economical, particularly in terms of energy conservation. In addition, most of the dust generated can be retained within the preheater, eliminating the majority of the dust problems.

Preheater systems do have difficulties when the raw feeds contain too high a percentage of alkalies, sulfur, or kerogen (combustibles). Several of the first

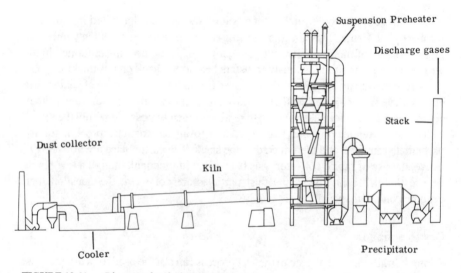

FIGURE 10-11.   Diagram of typical preheater system.

preheater plants in the United States were shut down in the early 1970s for one or more of these reasons. These problems do continue to plague the industry, though much progress had been made in recent years in dealing with them.[5]

Shaft kilns (Figures 10-12 and 10-13) constitute another possible dry processing technique, used to some extent in Europe. Shaft kilns have lower thermal and power requirements per ton of clinker produced than do rotary kilns,[6] being comparable to the preheater systems. Their major disadvantages are small capacity and a less uniform product, primarily as a result of tunneling of the gases through the load.

Semi-wet Processes

Semi-wet cement processing employs grate or grate-kiln methods (Figure 10-14). The grate-kiln systems use low strength, somewhat wet (approximately 12% water) 1-1 ½ cm pellets. These pellets are placed in a uniform bed on a travelling grate, hot air being blown upward from below. Dehydration and partial calcination occurs on the grate; the pellets are then fed to a short rotary kiln for the latter processing steps.

Advantages of grate-kiln systems include: 1) a controlled feed rate, 2) no flushing of materials into the kiln, 3) no segregation of raw materials due to differential shapes and densities, 4) avoidance of fluidization of the material bed, 5) very minimal dusting, 6) the production of uniform clinker, 7) relatively low energy requirements (70% of that required for a modern long kiln[7]), and 8) the

FIGURE 10-12.   Typical shaft kilns. (*Photo courtesy Wisconsin DNR.*)

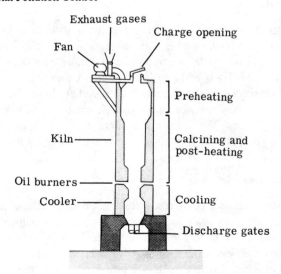

FIGURE 10-13.   Diagram of shaft kiln.

ability to use higher-alkali feeds than can many other processing techniques. The current difficulty encountered with grate processes is that they require nodules that are consistent in size and composition. This is often impossible to achieve with many raw materials using current technology.

Several of the newer installations in the United States do employ grate-kiln methods. Depending upon the local conditions, in some situations they are deemed preferable to preheater systems.

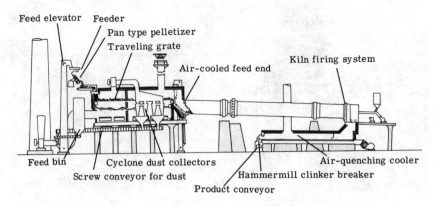

FIGURE 10-14.   Diagram of grate-kiln process.

ENVIRONMENTAL PROBLEMS[8]

The major environmental difficulty experienced by cement companies is the handling of the waste kiln dust that forms in the processing (Figure 10-15). This is particularly true for those plants employing wet processes. The water pollution problems that cement plants have are generally directly related to dust collection and/or dust disposal. For this reason, this discussion will focus on dust handling and disposal.

Kiln Dust Production and Characteristics— Wet Processes

The wet processes produce the largest quantities of dust, which must be collected and disposed of by some means. The majority of the dust is generated within the calcination region of the kiln; thus, chemically, it is composed of partially processed raw materials. Therefore, the dust particles are dry, with about 50% having undergone calcination (driving off of the CO_2), and a small percentage have at least started the clinkerization steps. In general, the slurry exhibits a potential CO_2 loss of 35–36% (by weight): the potential ignition loss for the dust varies between 13% and 35%. The smaller the loss, the more the dust resembles the clinker rather than

FIGURE 10-15.   Waste kiln dust can contaminate nearby streams, leading to serious pollution problems. (*Photo courtesy Wisconsin DNR.*)

the raw material. The dust tends to have a greater percentage of alkalies and sulfur trioxide (SO_3) than does the slurry—sometimes by a factor of 10 more. This does lead to difficulty in the reuse of some of these wastes. The chemical composition of typical cement dust is in Table 10-1.

The dust particles are usually less than 10 μm diameter. Many are less than 4 μm. The typical size varies with the source of raw materials and the exact processing, particularly the crushing and grinding methods to which they have been subjected.

The quantity of dust generated is dependent upon many factors. The exact type and size of the processing equipment, especially the chain system, the kiln air velocities, the type of fuel (coal-fired plants produce more dust), the chemical composition and the physical parameters associated with the raw materials, the percentage of water used to produce the slurry (more water reduces the dust production), and the size to which the raw materials are ground are all factors that affect dust production. Similarly, there have been many techniques suggested to reduce the amount generated. A few of these techniques include improving the chain systems, controlling air in-leakage, maintaining proper gas velocities, increasing the slurry moisture, using impactors rather than hammer-mills as the crushing technique, and not operating a kiln beyond its design capacity. Regardless of these suggestions, some plants nevertheless generate tremendous quantities of dust; some plants experience dust production rates up to 50% of the clinker weight. In these maximum cases, over one-third of the raw materials can react to form dust, not clinker. In a typical plant, the actual output of dust varies around 150–250 tons/day/kiln, depending upon the kiln capacity. Today, EPA guidelines demand that essentially 100% of this dust be collected. Usually, this is done by electrostatic precipitators. Environmentally and economically, it is far preferable to recycle this dust back into the process rather than to dispose of it—usually by landfill—at this stage.

TABLE 10-1   Composition of Typical Cement Dust

Compound	Percent
SiO_2	14.70
Al_2O_3	3.35
Fe_2O_3	2.18
CaO	47.20
MgO	2.10
SO_3	4.95
Na_2O_3	0.90
K_2O	4.80
Ignition loss (CO_2)	21.10

Markets for Waste Kiln Dust

The markets for this dust (as, for example, fertilizers) are limited, though for some soils it can be very beneficial. For example, in the early 1970s, one plant in Iowa was forced (like many others) by the EPA to cease emitting particulates from its stack. This dust had been, for many years, distributed over the nearby farmland. After the installation of much expensive collection equipment, the problem was solved—or so the plant thought. The farmers in the area, however, had appreciated the dust on their land, so they retaliated by suing the company, demanding that the company continue to allow the dust to be emitted. The farmers won. Thus, the company was ordered by the EPA to stop its emissions, and by the local courts to continue the emissions. The resolution of the controversy was that the company had to collect and bag some of the dust, and make it available free of charge to the local farmers.

In some recent tests, cement dust has been fed to cattle as a grain supplement, and the cattle appear to thrive. Some has also been used as a filler in road beds, as an aggregate in the production of cement blocks, and as a filler in plastic floor molding. Regardless of these possible markets, however, the majority must be either recycled or disposed of in landfill sites.

Landfill

Cement dust landfills are not of the "sanitary landfill" type. Usually, they are just old quarries or similar areas. The relative nonreactivity of these dusts does not demand dirt cover or similar precautions. The sheer volume of these wastes renders many such techniques impractical. Landfill procedures are costly in numerous respects. The dust has had a significant monetary and energy investment in its production, investments that would literally be thrown away if the dust were disposed of. Suitable locations for landfill sites near plants are getting more scarce as the quantities of waste grow. Many sites have environmental problems, such as leaching of the alkalies from the dust during rainstorms. The dust is of a very low density, and thus some sites can also be very dangerous: a person or animal who would accidentally fall into an area recently filled would sink and soon suffocate, in some locations. Proper handling and return of the dust to the system can save a typical wet process plant $50 or more per hour, and correspondingly, 1.4×10^{10} joules of energy/hour, for each kiln, plus the avoidance of these other related difficulties.[9]

Dust Return Methods

Because of the extremely large quantities of dust often formed in cement manufacture, it is very favorable for a plant to recycle its dust back into the process.

There are presently three common ways to reprocess the dust: 1) mixing the dust with the slurry feed, 2) pouring it into the kiln forward of the chain section by a system of scoops (Figure 10-16), and 3) insufflating, or blowing the dust into the kiln from the flame end, roughly parallel to the kiln axis (Figure 10-17). The simplest and most economical method of these is the first, mixing the dust with the slurry, but this technique has only limited applicability. Due to the chemical properties of the waste dust, and the partial processing that has occurred, if the dust comprises more than only a few percent of the total feed, the mixture partially sets and clogs the feed end of the kiln. Only a few cement companies can successfully employ this dust return process.

Scoop systems were very much in vogue during the 1960s. Most plants installed them on one or more of their kilns. Some installations, however, were never used. The technique has definite theoretical advantages, one of which is returning the dust to a location similar to where it was generated, thus providing a close matching of its chemical properties to those of normally processing materials, but many practical difficulties were experienced. Few plants currently use scoop systems because of the air in-leakage at scoop locations, leading to upset kilns, and also because of the recirculating dust loads often formed. The scoop systems pour the dust perpendicularly through the center of the kiln and thus through the turbulent air stream. Many of the fine dust particles become re-entrained and carried out again to the precipitator. Other problems include dust rings within the kiln and other maintenance problems.

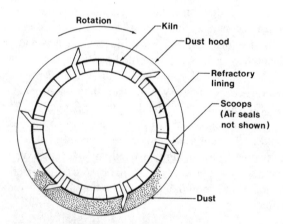

CROSS SECTION VIEW OF KILN
Dust from Precipitator fed to sleeve located
past chains; scooped and added to process
by kiln rotation.

FIGURE 10-16.   Scoop system for waste kiln dust return.

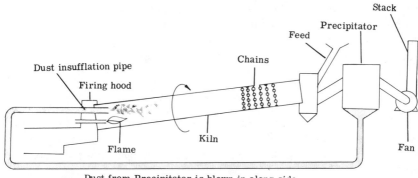

Dust from Precipitator is blown in along side
fuel pipe (gas, oil, or coal) or in fuel pipe (coal)

FIGURE 10-17.   Insufflation system for kiln dust return.

Insufflation is the most commonly used of these techniques and the most generally successful. The dust is blown into the kiln at 15–20 psi alongside the fuelpipe, or, in the case of coal, possibly along with the pulverized fuel. The fine dust heats very rapidly, and almost instantaneously becomes soft and sticky. The extreme turbulence of the flame allows for a relatively long residence time within the approximately 2700°F region of the flame, enhancing the probability of dust clinkerization, and also the likelihood of interparticle interactions, which lead to agglomeration and less dust carryout.

There are many factors that influence the succes of the insufflation procedure. For coal systems, some operators put the dust into the coal pipe, others use a separate insufflation pipe. Using a separate insufflation pipe seems to be more effective in the amount that can be insufflated, but it might not be as good for other process reasons. One effect of the insufflated dust is to cool and lengthen the flame. This has both advantages and disadvantages. On one hand, the flame temperature can be accurately controlled by varying the quantity of dust, and hot spots on the kiln walls can be eliminated or at least minimized. On the other hand, to be able to maintain an adequate temperature, more fuel must be used. Also, there is a limit to the amount of dust that can be used, because the permissible length of the flame is dependent upon the desired temperature gradients in the kiln. The most appropriate location of the insufflation pipe is thus dependent upon the chemical nature of the dust, its fineness, the kiln air velocity, and the desired amount of flame temperature control. The alkali content of the dust also limits, in the high-alkali situations, the amount of dust that can be returned.

One general limiting factor is the amount of control that a plant has over the dust as it is returned to the kiln. Accurate control—preferably by weight—is extremely important even under the best conditions. There appears, however, to be a maximum in the amount of dust that can be insufflated under ordinary conditions—an insufflated dust-to-clinker ratio of about 0.25. This maximum is likely due to the affect of the dust on the flame, and the recirculating loads that do develop. Some plants generate significantly more dust than this.

The General Portland, Inc., Signal Mountain Plant

Insufflation is totally unsuccessful in some plants—the recirculating loads that build up are just too large, even for a small amount of dust return. The General Portland, Inc. (GPI), Signal Mountain Plant in Chattanooga, Tennessee, is one such facility.

The Signal Mountain Plant has been located at the base of Signal Mountain (of Civil War fame), along the Tennessee River, for many years. Its raw materials are, in fact, barged in along that waterway.

In the early 1970s, the plant had experienced some air pollutant difficulties, aggravated by the fact that a very wealthy suburb was built in recent years on Signal Mountain itself, above the stacks of the plant. Many of these homeowners had taken GPI to court to sue for damages incurred by these buildings and the vegetation due to the cement dust. After a several-years-long court battle, this issue was settled, with compensations of $100 per individual being awarded to several of the homeowners. By the mid 1970s, however, the air pollution problems had been solved, the precipitators were operating efficiently, and the emissions were very low. At that time, however, they were experiencing another difficulty. Now that essentially all of the dust was collected, it had to be recycled or disposed of in some manner. For some inexplicable reason, insufflation was not successful at this particular plant. Instead, the dust was landfilled, until it was discovered that the landfill site polluted the nearby Tennessee River with high-alkali (high-pH) leachate during rainstorms. The EPA ordered the GPI to cease and desist this landfill practice. There were two options available to the company: to acquire another landfill site at a location at some distance from the river (and from the plant) or to develop another method of dust recycling. GPI chose to support the development and commercialization of a newly discovered technique for briquetting the waste kiln dust before its return.

Agglomeration of dusts, including cement dusts, is not new. Many methods have been devised, including low-pressure extrusion and pan pelletizing or drum pelletizing, and agglomeration by one method or another is becoming increasingly important in the disposal of many dusts. For the cement dust, however, these classical methods are unsatisfactory. The aim is to return the dust to the system, and to develop as small a recirculating load as possible. To accomplish this, the briquettes must be very hard. The classical methods all involve the use of water or some other type of binding agent, the use of which produces pellets with a low green strength. These pellets must be properly aged before they can be used, a process that is inconvenient at best. The use of a modified paired roll briquetting machine can eliminate these difficulties.

In a high-pressure briquetting machine, the dry particulate material is fed into the space between the two vertically-mounted paired rolls, the surfaces of which are recessed in cavities in a waffle iron effect (Figure 10-18). The material is compacted as the rolls come together. (Figures 10-19 and 10-20.) In order to produce high-strength pellets, the surfaces of the rolls must be coated with an oil such as a 10W40 oil. This oil becomes embedded in the briquette surface, forming a very hard, glazed crust. The briquettes so produced have sufficient mechanical strength to be fed into the rear of the kiln, along with the slurry, or fed by means of an already present scoop system, or, preferentially, they can be fired into the kiln from the front end, alongside the flame, a technique similar to insufflation.

This process was tested and was installed at the Signal Mountain plant. It shows promise of being a means for returning the dust to many plants such as this.

It is also possible that, in the future, this method can be used to process, by means of shaft or grate-kilns, the piles of dust that have been generated by some plants over the years and that constitute a large potential source of partially processed raw materials. The raw material reserve life of a plant can be extended considerably by recovering this dust, and this may, in the future, become essential in many locations.

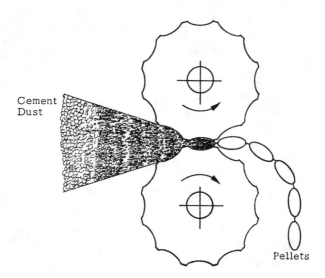

Cement
Dust

Pellets

FIGURE 10-18.   Formation of cement briquettes by high-pressure techniques.

FIGURE 10-19. Cement dust briquettes being formed.

FIGURE 10-20. Barrel of cement dust briquettes. (Note the quarter for size comparison.)

High-alkali Dusts

Many plants find that a portion of their dusts has too high an alkali content to be recycled. This dust can be fairly easily separated from the lower-alkali dusts right in the collection process. The alkalies vaporize more readily than do the other constituents of the dust; in fact, as the air stream enters the electrostatic precipitator, in some cases the alkalies may still be in the vapor phase. The alkalies, when they condense, also form extremely fine particles. For these reasons, the alkalies are more difficult to collect. If the plant uses a precipitator with several compartments, the alkalies tend to collect in the last compartment, the other dusts in the earlier compartments.

These high-alkali dusts can be recycled only if, by some means, the alkalies can be removed. The only process currently used, and that in only a few plants, is to leach out the alkalies in settling ponds. After leaching, the dust could be returned to the process. This procedure is not very popular. One of the most serious problems is the ultimate fate of the leaching water. It must eventually be treated for alkali removal. Another restriction is the large amount of land needed for these settling ponds.

Most plants must, instead, landfill their high-alkali dusts. This can create alkali leaching problems at the site, so proper location of these sites is very important.

Other Types of Kilns

Kilns of other than the wet process variety can have far fewer dust problems. High-alkali feeds (hence dusts) do create problems in preheater systems. These high-alkali fractions must be bled off from the process and then disposed of, probably by landfill. Grate kiln systems can handle even high-alkali feeds quite well, for the alkalies can be vaporized from the feed before it enters the kiln.

OTHER ENVIRONMENTAL CONSIDERATIONS

Only a few plants use wet scrubbers for dust collection, so water pollution at the plant is not generally a major concern. But as mentioned above, the leaching of landfill sites can be a serious problem, especially if the dusts have a high alkali content.

As with all mining operations, the quarrying of the cement raw material does create some environmental concerns. In particular, large quantities of $CaCO_3$ are utilized. Though there is no shortage of raw materials foreseeable in the future, the quarrying is usually very dusty and often noisy. Reclamation of old quarries is sometimes difficult, though in some locations they have been converted to recreational lakes.

The disposal of old concrete is another issue that must be considered. Its recycling is impossible due to the strong exothermic reaction that occurred during its formation. The materials can be crushed and used for fill in certain circumstances.

References

1. Ross, R. D., *Air Pollution and Industry*. New York: Van Nostrand Reinhold, 1972, pp. 2–3.
2. *Energy Conservation in the Cement Industry*, Conservation Paper No. 29. Washington, D.C.: Federal Energy Administration, #041-018-00095-0, 1976, p. 1.
3. *Ibid.* p. 2.
4. *Ibid.* p. 8.
5. Garrett, H. M. and Murray, J. A., "Improving Kiln Thermal Efficiency—Dry Process Kilns, Part 4." *Rock Products* **9**, *No. 58*, 1974, p. 9.
6. Peray, K. E. and Waddell, J. J., *The Rotary Cement Kiln*. New York: Chemical Publishing Company, Inc., 1972, p. 1.
7. Allis-Chalmers, Folder 07B8431, 1974.
8. Sell, N. J. and Fischbach, F. A., *Dust Return to Rotary Cement Kilns*. Report for Allis-Chalmers, November 27, 1974.
9. Sell, N. J. and Fischbach, F. A., "The Economic and Energy Costs of Dust Handling in the Cement Industry." *Resource Recovery and Conservation* **3**, *No. 4*, 1979, p. 468.

11

Glass Manufacture

Man has been manufacturing glass since sometime between 10,000 and 3000 B.C. Thus, this is one of our oldest industries. The first factory in the United States was, in fact, a glass plant, built in 1608 at Jamestown, Virginia. The early glassmaking processes were very crude, and the procedures slow, laborious, and costly. Until the invention of the blowpipe in 300 B.C., glass was purely a luxury item. Today, however, there are 350–400 glass plants in the United States alone. Every individual uses many glass items every day—items ranging from eyeglasses and cameras to glass jars and windows.

THE COMPOSITION OF GLASS

A glass is, scientifically, a noncrystalline (amorphous) solid. It forms because the viscosity of the molten glass is so high that, as the temperature is lowered, the constituent molecules lose their translational kinetic energy before they can migrate to their proper positions to form crystals.

All commercial glasses are based on silica (SiO_2) available in the form of sand. In addition to silica, boric oxide (B_2O_3), aluminum oxide (Al_2O_3) or alumina, and phosphorus pentoxide (P_2O_5) are major constituents of glass. Other constituents include the alkaline oxides; sodium oxide (Na_2O) or soda ash, potassium oxide (K_2O), and lithium oxide (Li_2O); and the alkaline earth oxides, calcium oxide (CaO) or lime, and magnesium oxide (MgO) or magnesia. Lead oxide (PbO) is important in certain high-quality, dense glasses; iron oxides or similar substances can be used for color.

About 90% of today's glass is "soda-lime" glass. It is used for plate and window glass, most containers, electric light bulbs, and many other industrial objects. The composition is about 72% SiO_2, 15% NaO_2, 9% CaO, and the remaining 4% various minor constituents.

"Soda-lead," or lead glass, employs PbO instead of CaO. Often, the SiO_2 content is simultaneously reduced, resulting in a glass that is as much as 30% PbO. Lead glass is much more expensive than soda-lime, but it is soft, easy to melt, and has excellent optical and electrical properties. This type of glass is used for the finest tableware and art objects, for optical purposes, and in certain electrical applications.

Borosilicate glass, commonly called pyrex, is about 80% SiO_2, 4% total alkali, 2% Al_2O_3, and 13% B_2O_3. This composition provides a glass that is very heat/shock-resistant due to its low thermal expansion coefficient, about one-third that of soda-lime glass. It is used for chemical and electrical purposes, in baking ware, and in glass pipelines.

Aluminosilicate glasses, those containing Al_2O_3, are also heat/shock-resistant. In addition, they can withstand higher temperatures than can the borosilicate glasses. Thus, ovenware is usually made of borosilicates, but the glassware designed for cooking on an open flame on top of the stove is made of aluminosilicates.

Another type of glass is "96% Silica Glass," or Vycor. It can withstand being heated red-hot and then plunged into ice water without cracking. It is manufactured from a borosilicate composition that is first treated to leave a porous skeleton. When heated, it shrinks 35%. This glass is mainly used for chemical supplies.

THE RAW MATERIALS[1]

Silica sand, of appropriate composition to produce glass, is derived from quartz sandstone. It consists of over 99% SiO_2, with the impurities typically being clay and zircon. The major sources are former shallow marine environments, often located far from the glass plants. Transportation costs, therefore, are substantial. Fortunately, there appears to be no shortage of SiO_2 in the forseeable future. Fused Si_2, obtained by melting pure sand, is a good glass for some purposes but is difficult to make because it requires very high temperatures to melt (1723°C), and it remains very viscous even at these high temperatures.

Na_2O is added primarily to act as a flux, to lower the melting point. It is generally manufactured from NaCl, ordinary table salt. The salt is mined by first pumping water into wells, followed by pumping out the brine.

Most Na_2O is produced by the "Solvay" process. The salt brine is mixed with NH_3 in the presence of CO_2. The result is the formation of ammonia bicarbonate (NH_4HCO_3), followed by the formation of ammonium chloride (NH_4Cl) and sodium bicarbonate ($NaHCO_3$):

$$H_2O + NH_3 + NaCl + CO_2 \rightarrow NH_4HCO_3 + Na^+ + Cl^- \rightarrow NH_4Cl + NaHCO_3.$$

The $NaHCO_3$ is then calcined to Na_2O:

$$2NaHCO_3 \rightarrow Na_2O + 2CO_2 + H_2O_1.$$

The NH_3 can be regenerated from the NH_4Cl by using calcium hydroxide (CaOH):

$$2NH_4Cl + Ca(OH)_2 \rightarrow 2NH_3 + CaCl_2 + 2H_2O.$$

The resulting $CaCl_2$ must be collected and disposed of. This is currently a major environmental problem with many soda ash plants.

There has recently been an increase in the use of naturally occurring Na_2O. The major direct source of Na2O is trona rock deposits, consisting of 90% sodium sesquicarbonate (Na_2CO_3 · $NaHCO_3$ · $2H_2O$) and 10% insoluble impurities. The trona can be dissolved, the impurities removed, and $NaHCO_3$ recrystallized from solution. This $NaHCO_3$ can then be calcined to 99.9% pure Na_2O. Na_2O can also be obtained directly from naturally occurring sodium carbonate (Na_2CO_3), which can be calcined appropriately.

The natural supplies of soda ash raw materials are large, yet in recent years there was a shortage, generally attributed to the tightening of water discharge restrictions for the Solvay plants. This shortage has been blamed for the shortage of canning jars experienced in the early 1970s. Currently, about 55% of the total U.S. soda ash market is for glass manufacture.[2]

Glasses containing just Na_2O and SiO_2 are usually low-durability and easily attacked by water. The addition of CaO and/or MgO (often from the mineral dolomite), and even some Al_2O_3, greatly improves the durability of the glass. These substances are likewise generally obtained by mining, though the small amounts necessary can also be provided by waste products from other industries.

THE MANUFACTURING PROCESS

The raw materials are transported to the plant, generally by railroad tank car, and stored as dry powders in silos. The raw materials are weighed and then mechanically mixed in the proper proportions. To the mix is added 20–40% cullet, or waste glass, from a previous melt of the same type of glass. Not only is this a resource recovery procedure, but it also assists the melting. In recent years, pelletization of the mixed dry powders by pan pelletizers to form soft, semi-wet pellets has sometimes been done, also to assist in the melting. The materials then are transported to the melting units by "batch cars," hoppers, rail-driven buckets, or conveyor belts.

The melting unit varies with the type of glass. Optical glass, rolled plate glass, and art glass are typically made in refractory pots, which can process up to 3000 lb of glass in a batch-type unit. Larger quantities of glass are melted in furnaces called "day tanks," which are batch units for up to 4 tons of materials, or in "continuous tanks," which can process up to 1000 tons at once. In a continuous tank process, the weighed raw materials are fed intermittently into the top of one end. Flames, typically employing natural gas as fuel, melt the materials. The maximum temperature required is about 1550°C. As the materials melt, the impurities sink and can be removed. The molten glass, often 5 ft deep and up to 125 ft long, is drawn off from near the bottom of the tank at the "throat" or "skimmer." Residence times may range from 24 to 48 hours.

Within the furnace, various processes are occurring.[3] First the free water in the raw materials evaporates. At about 600°C, the ingredients react to form sodium calcium carbonate; this further reacts with SiO_2 at a slightly higher temperature, emitting CO_2. By the time the mix reaches 780°C, the mixture of Na_2CO_3 and sodium calcium carbonate melts and flows down the tank, initiating many various reactions. SO_2, SO_3, nitrogen oxides, and bound water are vaporized, as are Na_2O, B_2O_3, and fluorine gas (F_2). The SiO_2 reacts with CaO to form $CaSiO_3$, which then further reacts to form the final glass compounds.

Near the bottom of the furnace, the temperature reaches about 1550°C, depending upon the type of glass. The molten glass is less viscous at these temperatures, and the gas bubbles rise toward the surface. A rapid melt produces fewer, larger bubbles, which rise rapidly, simultaneously helping to mix the molten materials. A slower melting leads to the formation of small bubbles, "seeds," which are difficult to remove and lower the quality of the product. Often, seeds can be removed by chemical additives, called fining agents. These additives can act by: 1) reducing the viscosity and surface tension, 2) reacting with the gases in the seeds, and/or 3) volatilizing in the melt, thus promoting mixing.

The molten glass flows through the "throat," and is cooled to approximate working temperatures. The materials will often have been in the furnace for up to five days.

The glass is shaped by one of several means. It can be drawn to form rods, tubes, and flat glass. It can be blown by hand. Or it can be shaped by machines—pressed, blown into molds, or cast.

It is critical to control the cooling rate carefully once the glass becomes rigid, or stresses develop due to different cooling rates occurring in a different parts of the object. Controlling the cooling so as to avoid such problems is annealing. Often the glass is reheated, and then cooled at a definite rate. Because of the low thermal conductivity of glass, uneven cooling cannot be avoided. As glass cools, it contracts due to a decrease in the average intermolecular distance: the SiO_2/O_2 network rearranges, the density increases, and the object shrinks. If the glass does not crack as it is cooled very slowly, and an even temperature is reached, no permanent stresses form. However, if the glass cools rapidly, it shrinks less (in total) than if cooled slowly. Thus, if one part cools much more rapidly than another part, permanent (weakening) stresses can develop.

It is also possible to take advantage of such stresses to strengthen the glass. This is done in tempering. If the surface is cooled quickly, it shrinks less than the interior. The interior is thus prevented from shrinking fully. The net result is tension in the interior and compression on the surface. Surface cracks must overcome this compression in order to spread, and thus the glass is strengthened.

The glass can then be further treated by glazing, polishing, and/or coating.

The processing of molten lead glass to form decorative "lead crystal" tableware is shown in Figures 11-1 to 11-8. Much of this type of glass is hand-blown. The

FIGURE 11-1. The molten glass, to be blown or pressed into a decorative item, is removed from a furnace after a residence time of 24 to 48 h or more, as shown here at the Rossi Glassworks in Ontario, Canada.

FIGURE 11-2. A gatherer (left front) collects a glob of glass on the end of a blowpipe at Cavan glassworks in Cavan, Ireland.

FIGURE 11-3.　A blower blows the glass into the desired shape.

FIGURE 11-4a.　A gaffer a) left front and b) top, next page, does the final shaping of a blown object.

FIGURE 11-4b. (*Continued*)

FIGURE 11-5. Molten glass being pressed into a mold to produce, in this case, a fluted cranberry glass bowl.

FIGURE 11-6.   Marking a glass for correct pattern cutting.

glass must be maintained at about 1000°C in order to be soft enough to work. Figures 11-2 to 11-4 illustrate several stages of the blowing operation: "a gatherer" collects a glob of glass on the end of the blowpipe; a "blower" blows the glass into the desired shape; a "servitor" shapes a stem and/or base to be added to the glass object; then a "gaffer" does the final shaping and finishing.

Figure 11-5 illustrates another technique; forming the glass by a sand mold.

The crystal next must be marked according to the desired pattern (Figure 11-6). The patterns are cut deep into the glass (Figure 11-7) by sandstone or carborundum wheels. The object must then undergo a final grinding (Figure 11-8), using very fine abrasives. Finally, the glass can be chemically etched by hydrofluoric acid to better restore the original luster.

THE ENVIRONMENTAL IMPACTS OF GLASS MANUFACTURE[4]

The most serious environmental problems of the glass industry are related to air and water pollution. Within the industry, solid wastes are not a major difficulty. The trimmings, rejects, and similar materials can be recycled as cullet.

FIGURE 11-7.   Cutting the correct pattern into glass bowls.

Water pollution is a problem in a variety of processes. At the initial mining step, large quantities of water are required for beneficiation of the sand by screening and washing, for leaching with acid, flotation, and other processing techniques. In fact, more water is required for sand and gravel mining than for any other mineral industry. This not only results in a lowering of the groundwater tables, but also in environmental problems due to runoff.

The Solvay process for the manufacture of soda ash creates large quantities of a calcium chloride (saline) effluent. The recent stricter regulations on this effluent could not be met by some plants, forcing them to close. It appears that the Solvay process will soon be replaced by the mining of naturally occurring deposits, a procedure with fewer environmental difficulties.

Water pollution is also experienced via the runoff from cullet piles. Much of the cullet is impregnated with lubricating oils, which can be leached away.

Other water pollution problems arise because of the need to use special manufacturing techniques for specialty glasses. For example, in the manufacturing of eyeglasses, acetone is used as a solvent, acrylics for surface finishing, and other organic liquids for treating the glass. These liquids must be treated by (for example) a pyro-decomposer, to convert them to harmless substances.

The pollutants involved in the actual production of glass, the melting and

FIGURE 11-8.   Final grinding with very fine abrasives.

forming, are primarily particulates and various gases, especially sulfur oxides and nitrogen oxides.

As the raw materials are initially fed into the furnace, the volatile materials in the feed and in the fuel are released, and other noxious gases, such as the nitrogen oxides and F_2, are formed. The hot escaping gases entrain the alkali oxides and sulfur oxides, which are vaporized and then carried out of the system. "Checkers," or brick grids, are placed in the path of the exiting air stream. These checkers both help to recover heat and aid in pollution control.

The particulates generated within the furnace are a major problem, due to their small size, which is usually 1–30 μm in diameter, depending upon the type of glass. In general, the greater the production rate, the higher the temperature required, and hence the greater the fuel consumption and particulate load.

Most of the dust problems are created when the particulate matter from the processing raw materials is carried out with the emitted hot gases.

Particulates are also formed due to the fuel combustion. Natural gas is relatively clean, but various fuel oils can generate a very opaque plume. The lubricating oil on the cullet (present due to lubrication of the glass-forming molds) can be partially combusted and flash volatilized, producing smoke.

The composition of the particulates varies with the type of glass, the raw materials, and the furnace, but typically condensed oxides (Na_2O and K_2O) comprise 25-40%, SO_3, 25-50%; CaO, 1-20%; and As_2O_3, 0-8%. There are also present smaller amounts of PbO, Al_2O_3, Fe_2O_3, MgO, ZnO, and F_2. Generally, about 2 lb of particulates are produced per ton of glass.

The quantity of particulates formed can be reduced by increasing the checker volume, the area of the melter, and the percent of cullet used. The cullet helps the finer materials melt without itself volatilizing, though there appear to be difficulties in seed formation if the percent composition is too large. Increasing the raw material particle size also minimizes the particulate formation, but increases the fuel required for melting.

The very small particulates create a very visible plume due to light diffraction. Collection is difficult, for electrostatic precipitators are not always successful. High-energy scrubbers are one possibility, but they require significant water and energy, and lead to water pollution problems. Baghouses are another possibility; tests as to appropriate fabrics are being made.

Though the glass industry does not produce as many particulates as do some other industries, the difficulty it has collecting the fine particulates produced at high temperatures is particularly significant.

A certain amount of particulate matter also becomes airborne during the raw materials transport and unloading operations. This is, in general, insignificant due to the larger particle size, about $300\mu m$ in diameter on the average. These particles quickly settle, and never develop into a major problem.

THE USE OF WASTE PRODUCTS IN GLASS MANUFACTURE

Though solid waste is not a significant problem within the glass industry itself, the large amount of waste glass produced annually, particularly glass containers, is a major municipal solid waste problem. There is much opportunity for this waste glass to be crushed and used as cullet within the glass manufacturing process (Figure 11-9). There are, however, technical as well as societal difficulties. As mentioned, the increased use of cullet may lead to "seedy" products, whose rejection by the public is feared. Impure cullet can create problems in the furnace, particularly if the impurities are not constant and predictable. Differences in viscosity with the mix can also be a problem. In addition, separation from the municipal refuse has economic disadvantages.

Another potential glass raw material is waste blast furnace slag. The Al_2O_3, SiO_2, and CaO contents are useful, and the free iron can be separated out magnetically. A good quality glass has been made, on an experimental basis, with up to 31% slag.

Again, there are several problems to deter this process from ever becoming widespread. The sulfur in the slag would increase the air pollution, and unless it

FIGURE 11-9.   Separation of glass and cans for recycling. (*Photo courtesy Wisconsin DNR.*)

can be eliminated from the product (for example, by the addition of arsenic to produce arsenic sulfide in the flue gases), only a black glass is possible. The glass is also not particularly suitable for blowing—only rolling or pressing. In addition, the slag is often of variable composition, and, in most cases, the sources are not located near the glass plants.

In spite of these difficulties, there is evidence that there will be increased use of slag and other materials in the near future as raw material and energy costs rise.

NEW MANUFACTURING PROCESSES[5]

Due to both high energy costs and increasingly stringent air pollution regulations, the glass industry has been developing new process technology. Current technology using old, inefficient reverberatory furnaces with massive melt chambers requires large amounts of fuel to maintain proper melt temperatures. In addition, the typical heating process—firing a mixture of natural gas and air over the molten glass— produces significant NO_x and particulates.

As a partial response, some plants have installed electric melting, especially for the smaller facilities. For larger operations, it is used only for supplemental heating. Over 80% of the U.S.-produced glass is still melted in gas-fired furnaces.

A number of new processes also are being developed, with the goals of increasing energy efficiency and lowering NO_x, SO_x, and particulate emissions. Most of these processes promote rapid melting by rapid flow, violent mixing, and convective heat transfer. Methods under investigation are designed to reduce heating times to about 30 seconds, compared to the several hours to 1 day common with conventional technology. Most of these initial studies have been designed for fiberglass production, due to its increased tolerance for seeds—about 100 seeds per mL, or 100 times that allowed for flat glass.

Pure O_2 has replaced combustion air in a few commercial applications. Pure O_2 burns without significant NO_x production, results in substantially fewer particulates, and can lower the natural gas demand by up to 25%. However, it is more expensive and can cause hot spots in the melt, possibly damaging the refractory liners. It is also more reactive and hence more difficult to preheat than air.

Refining of plate- and container-glass normally requires 30 hours, due to the time necessary to remove the bubbles from the massive amounts of molten material. Attempts are on-going to also reduce this time to about 1 hour, by use of thin glass films.

The use of cullet is also increasing, as more glass containers are recycled. In 1989, about 750,000 tons of glass were recycled in the United States, comprising about 20% of the feed for container glasses. In some parts of Europe, as much as 60% of container glass can be of recycled material. It is estimated that the amount of cullet in a given feedstock will more than double in the United States by the mid-1990s. Not only will this decrease the load on landfills, but cullet requires less energy to melt than does virgin glass, and preheating it (by, for example, waste heat in combustion gases) can further improve furnace efficiency.

OTHER USES OF WASTE GLASS

There have been many studies as to how to use waste glass in other products. For example, waste glass can substitute for rock and/or sand as aggregate in asphaltic concrete, forming "glasphalt." Crushed glass can be mixed with clay and water and fired to form wall panels. The glass can be crushed, then mixed with water, calcium carbonate, and bentotine, and sintered. The internal vapors expand, forming a foamed glass. Crushed glass can be used as an aggregate in "slurry-seal," a pavement sealant that protects against moisture, provides skid resistance, or can simply be used to fill cracks and potholes. Glass wool, used for heat or sound insulation, can be manufactured from waste cullet. Waste glass can take the place of marble chips in terrazzo flooring, or it can be used to produce building bricks.

In spite of these many possibilities, the markets are not yet good for waste glass, and the waste glass quantities are climbing. Even if large quantities were utilized, this would not significantly decrease the glass solid waste problem. More changes are needed, such as an increased use of returnable bottles, or even the development of a degradable glass container.

References
1. Samtur, Harold R., *Glass Recycling and Reuse.* Madison, Wisconsin: Institute for Environmental Studies.
2. "Key Chemicals: Soda Ash." *Chemical and Engineering News,* February 26, 1979, p. 14.
3. Samtur. p. 28.
4. *Ibid.* p. 29.
5. Parkinson, Gerald, "Glassmaking Responds to New Challenges." *Chem. Eng.* **97** *No.10,* October 1990, pp. 30-37.

12

Paper and Pulp Industry

The paper industry is a very large and growing segment of the U.S. economy. As of 1990, 394 companies operated 603 paper mills and 351 pulp mills at 798 locations; they produced 76 million tons of paper products yearly and employed almost 700,000 people.[1] The U.S. consumption is also the greatest in the world. In 1988, per capita use of paper and board averaged 698 lb, almost 2 lb/person/day.

To get a better idea of the size of the paper industry, one can consider International Paper, the largest company in the industry. It ranks second only to the federal government in total acreage owned across the nation. In the state of Maine, paper companies control about 52% of the total land area. Overall, the estimated total land holdings of all pulp and paper companies amount to 50 million acres, an area equal to all of New England plus New Jersey.[2]

Paper is composed primarily of cellulose. The purpose of the pulping procedure is to separate the noncellulose components from the raw materials, leaving behind almost pure cellulose, to which other substances can be added to develop special qualities if so desired.

Any fibrous material is suitable as a raw material. In the United States, in practice, wood (both coniferous and hardwoods) provides almost 77%, waste paper 23%, and rags and other plant materials less than 1% of the remaining raw stocks.

Wood is composed of cellulose, other carbohydrates such as starches and sugars, and lignin, which acts as an adhesive for cellulose. Dry wood is typically 45% cellulose, 21-25% lignin, 25-35% hemicellulose, and 2-8% extractives (resin acids, phenols, fatty acids, terpenes), by weight. In plants, the cellulose determines the nature of the fiber and permits its use in papermaking. Cellulose is a polymer of glucose. Its chemical formula is $(C_6H_{10}O_5)_n$, where n is the number of repeating sugar units. Though n varies with the source of the cellulose and the processing it has undergone, typical values are 600-1500.

FIGURE 12-1.   A paper machine. (*Photo courtesy Wisconsin Paper Council.*)

The structure of cellulose is illustrated in Figure 12-2. Within plant materials, the cellulose molecules are oriented in long, parallel chains, to form fibrils. The fibrils, in turn, coalesce to produce the basis of the wood fibers. Some portions of the cellulose fibril are crystalline, while other portions are amorphous. Cellulose is insoluble and stable in basic solutions, an important fact when considering pulping.

Hemicelluloses are also sugar polymers, based on monomers of glucose, mannose, galactose, xylose, or arabinose. The hemicelluloses are found associated with the cellulose or with the lignin. The hemicelluloses are much less stable than cellulose and so readily degrade in the pulping process.

Lignin (Figure 12-3) is a very complex polymer, with a structure that is largely not completely understood. It also varies with the species of tree. It is known to consist of aromatic rings (carbon rings with alternating single and

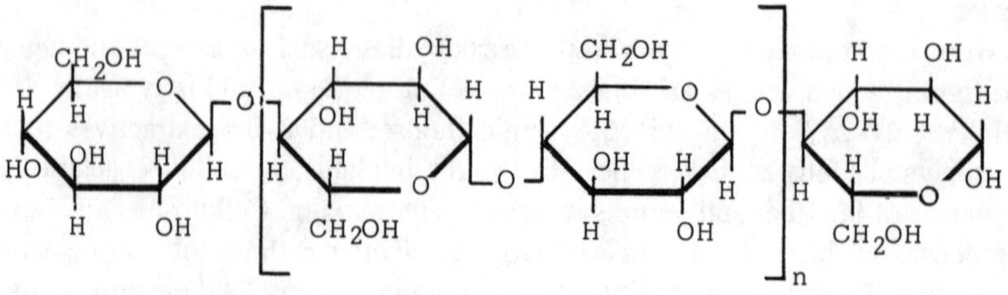

FIGURE 12-2.   Cellulose molecular structure.

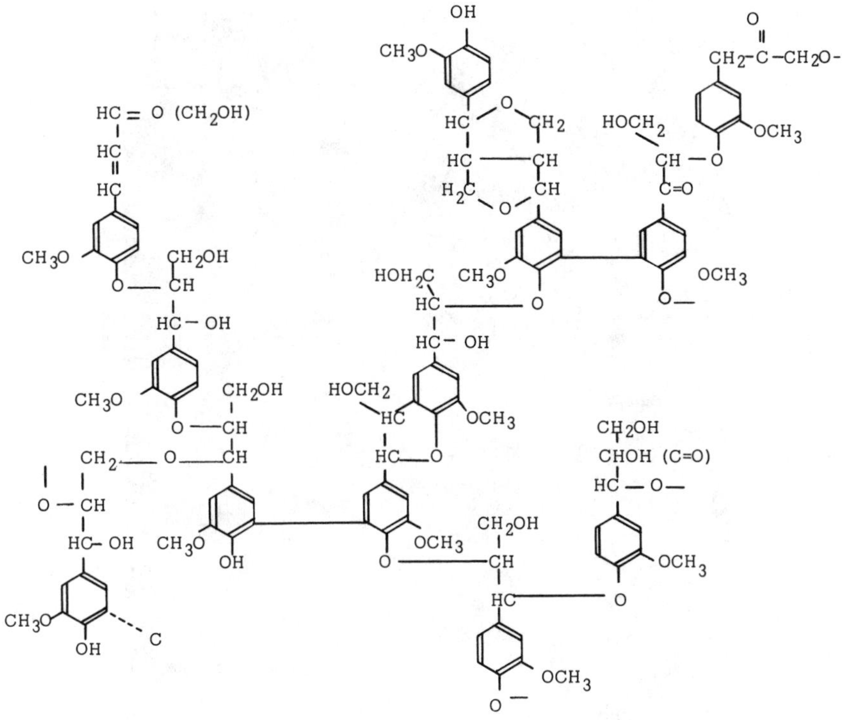

FIGURE 12-3.   Postulated lignin molecular structure.

double bonds), with attached hydroxyl (OH$^-$) and methoxyl (CH$_3$O$^-$) groups. It is found mainly between the cellulose fibers, and serves as the glue to bind them together.

Extractives are a diverse group of substances that are mostly soluble in water or organic solvents. Southern pines have a relatively high content, permitting the recovery of substantial amounts of tall oil and turpentine as an alkaline pulping byproduct.

The raw materials are first transported to the pulping facility (Figures 12-4 and 12-5) where they are stored for subsequent use (Figure 12-6).

The papermaking process consists of four basic steps:

1. Breaking down the raw materials (the wood, usually) to fine particles by mechanical means.
2. Dissolving the lignin and other noncellulose compounds chemically so as to form pulp.
3. Forming a layer of pulp slurry, which is then dried on a paper machine, yielding a mat of cellulose fibers, a paper sheet.

FIGURE 12-4.   The main ingredient of paper is wood, both hardwoods and softwoods. In this case, the trees are those available in Germany's Black Forest.

4. Possible addition, at some point in the process, of dyes, coating materials, or preservatives, depending upon the end use of the paper.

The broad environmental pollution problems associated with paper production are three-fold:[3]

1. Air pollution problems arise in the conversion of wood to paper pulp.

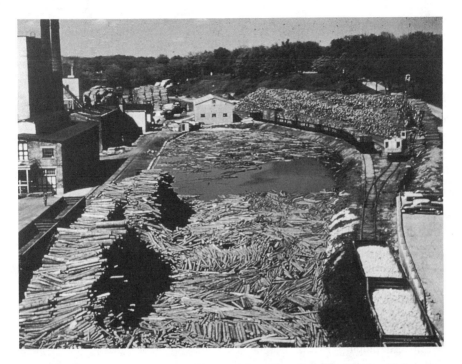

FIGURE 12-5. The wood is transported by waterway, truck, and/or rail to the pulp mill site.

FIGURE 12-6. Stacking the lumber at the paper mill. (*Photo courtesy Wisconsin Paper Council.*)

257

2. The pulping process releases much BOD and possibly trace amounts of toxic substance.
3. The conversion of pulp to paper sheet requires large amounts of energy, whose production generates air pollution. Additional BOD and some specific chemicals are also released in this process.

Figure 12-7 diagrams the overall papermaking process.

Of these steps, the pulping step is the one of most concern when discussing pollution. The drying of the slurry is the most energy-intensive step, however; thus, it also has serious environmental effects.

PULPING PROCESSES

There are three general types of pulping processes: chemical pulping, semichemical pulping, and mechanical pulping methods.

Mechanical Pulping

Mechanical pulping (10% of the total U.S. pulp production) uses only mechanical forces to separate the individual wood fibers. The traditional method is the groundwood process, where a block of wood, typically a whole debarked log (Figure 12-8) is pressed lengthwise against a rotating grinding stone (Figures 12-9 and 12-10). The fibers are torn from the wood, abraded, and then suspended in water. Since the fibers are short, the resulting paper sheets are of very low strength.

A more recent development in mechanical pulping is refiner mechanical pulp (RMP). The wood is chipped and then shredded and ground between the rotating disks of a refiner. These refiner disks are designed with a series of different sized and spaced radial bars that "unravel" the fibers in the wood chips. The RMP fibers tend to be longer than groundwood fibers, and produce a stronger paper sheet.

A further advance based on this process is thermomechanical pulping (TMP), in which the wood is steamed prior to and during refining. This pulp exhibits even greater strength characteristics than does RMP pulp.

In the production of mechanical pulp, essentially all of the wood components, including the lignin, remain within the pulp. As a result, the impurities can cause discoloration and the paper yellows upon aging. This pulp is, however, preferred for very short-lived uses, such as newspapers, where neither strength nor surface quality are important and low cost is critical. One major advantage is the yield: Up to 95% of the wood components (on a dry basis) are converted into pulp. However, it requires a tremendous amount of energy, particularly electrical power, to accomplish this.

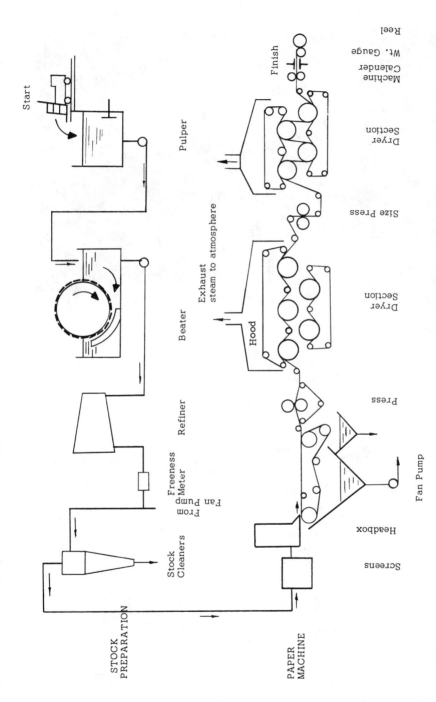

FIGURE 12-7. The overall papermaking process.

259

FIGURE 12-8. Debarking of the wood. (*Photo courtesy Wisconsin Paper Council.*)

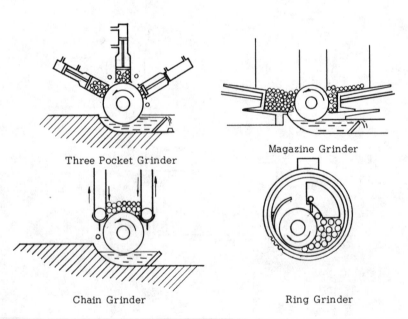

Three Pocket Grinder

Magazine Grinder

Chain Grinder

Ring Grinder

FIGURE 12-9. Schematic of typical mechanical pulpers.

FIGURE 12-10.   Typical 3-pocket grinders for production of groundwood pulp.

Chemical Pulping

Chemical pulping methods, which include primarily kraft (78%) and sulfite (3%) processes, utilize significant amounts of chemicals, along with heat and pressure, to solubilize the lignin and, incidentally, much of the hemicellulose. These processes dissolve a good portion of the wood material; yields are typically 40-50%.

All of the chemical pulping methods start with wood chips (Figure 12-11). The overall pulping process is illustrated in Figure 12-12.

Soda Pulping

The soda process is the oldest of the chemical pulping processes, and is not widespread at this time. It uses sodium hydroxide (NaOH) as the base at a pH of about 13-14. The solution is thus 0.1-1.0M in OH^- ions. A little sodium carbonate (Na_2CO_3) is also added. The coarsely ground raw materials are cooked in a closed digester (3-4 stories tall) with the NaOH solution at 170°C, 100-110 psi of pressure, for 3-8 hours. During this cooking time, the pressure is occasionally relieved by releasing the processing gas, mainly CO_2, to the atmosphere. The lignin polymer segments are dissolved by means of proton (H^+ ion) transfer from the phenolic OH—groups of the lignin to the OH^- ions available in the solution:

FIGURE 12-11. Wood chips that are to be chemically pulped. (*Photo courtesy Wisconsin Paper Council.*)

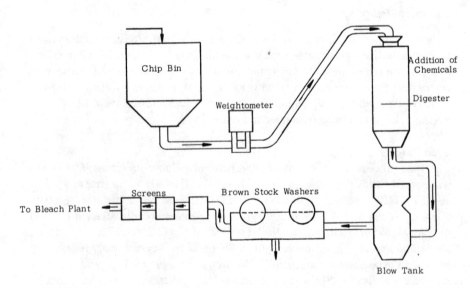

FIGURE 12-12. The chemical pulping process.

262

The other water-soluble substances simultaneously dissolve, leaving the cellulose pulp. The remaining liquor is recycled to recover the NaOH, which is relatively expensive. This is accomplished by first evaporating the water. Left are solids consisting of organic residues and inorganic sodium salts. Most of the NaOH originally present reacted with either the lignin or the CO_2 during cooking. The latter reaction produces Na_2CO_3:

$$2Na^+ + 2OH^- + CO_2 \rightarrow Na_2CO_3^- + H_2O.$$

The residues can be ignited; the organics will burn away, and the NaOH can then be recovered to a large extent by treatment with lime (CaO). This is called "recausticization," and proceeds as follows:

$$CaO + H_2O \longrightarrow Ca(OH)_2$$
$$Ca(OH)_2 + Na_2CO_3 \longrightarrow CaCO_3 + 2NaOH$$
$$CaCO_3 \longrightarrow CaO + CO_2.$$

Kraft Process

The soda process has been largely replaced by the sulfate (or kraft) process. This process includes not only NaOH, but also sodium sulfide (Na_2S), in the cooking liquor. The purpose of the sulfide is to provide a reserve of base, so the reaction can occur at a more uniform rate.

In water solution, the sulfide ion (S^{2-} hydrolyzes to form hydroxide ions (OH^-) and hydrogen sulfide ions (HS^-):

$$S^{2-} + H_2O \rightleftharpoons HS^- + OH^-.$$

The initial high concentration of NaOH (hence OH^- ions) forces the equilibrium to the left, according to LeChatelier's principle. However, as the OH^- ion is consumed, the equilibrium shifts to the right, replacing some of the OH^- ions. The net result is that delignification occurs at a more steady rate, not initially very fast and then later much more slowly. An additional advantage of this process is that the HS^- can also react with the lignin to enhance its solubility. The residual chemicals are very dark; hence, this is frequently called the "black liquor" process.

As with the soda process, gases must be occasionally released from the digester (Figure 12-13) during processing. In this case, however, the gases contain many sulfur compounds, which can generate air pollution problems. Figure 12-14 is an electron microscope picture of the resulting pulp.

Chemical recovery (Figure 12-15) is an integral part of the kraft pulping process. The spent liquor (black liquor) is first concentrated to about 50% solids in multiple effect evaporators. A second concentration stage, to at least 60% solids, is then conducted in concentrators or equivalent. This heavy black liquor is next inciner-ated in a recovery furnace under reducing conditions (that is, with a shortage of oxygen). The sulfide ion (S^{2-}) had largely been oxidized to sulfate (SO_4^{2-}) during the pulping; burning the dissolved organics in the black liquor under reducing conditions converts the sulfate back to sulfide:

$$SO_4^{2-} + 2C \rightarrow S^{2-} + 2CO_2$$

FIGURE 12-13.   A 1000 t/day capacity continuous kraft pulping digester located at Metsa-Botnia, Finland.

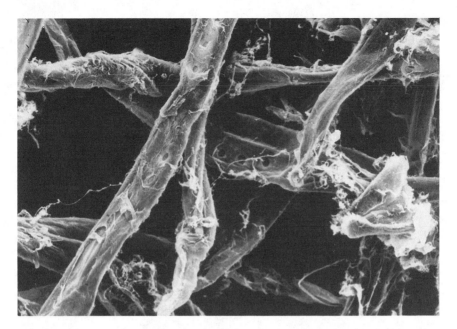

FIGURE 12-14. Picture of pulp fibers as observed through a microscope. (*Photo courtesy Institute of Paper Chemistry.*)

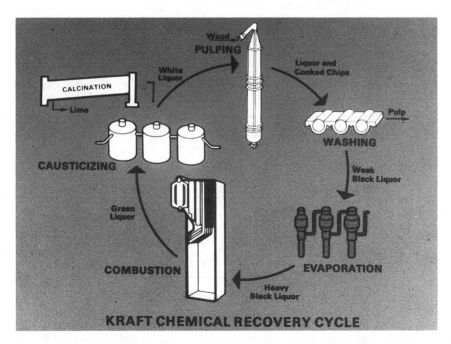

FIGURE 12-15. Schematic of the overall kraft spent liquor recovery process. (*Photo courtesy of David Clay.*)

Sufficient oxygen is available to oxidize the organics to primarily CO_2 and H_2O. The "ash" at the bottom of the recovery furnace, called smelt, is a mixture of primarily Na_2S and $NaCO_3$. This smelt is then dissolved in water to form "green liquor." The green liquor is reacted with lime ("causticized") to regenerate NaOH. The resulting white liquor is a mixture of primarily sodium sulfide and sodium hydroxide, which can again be used for additional pulping.

The caustization creates another product: insoluble calcium carbonate ($CaCO_3$). This, too, is recycled by heating in a lime kiln to regenerate lime for further processing.

Sulfite Process

The sulfite process employs the hydrogen sulfite (bisulfite) ion (HSO_3^-) to dissolve the lignin rather than using NaOH. The HSO_3^- reacts in the digester with the phenolic OH—group on the lignin, forming a sulfonic acid:

The net result is to increase the solubility of the lignin in the solution.

The ion can be formed initially from sulfur. The sulfur is oxidized ($S + O_2 \rightarrow SO_2$) and then hydrated ($SO_2 + H_2O \rightarrow H_2O_3$). When the solution is made more basic, the HSO_3^- ion is generated: $H_2SO_3 + OH^- \rightarrow HSO_3^- + H_2O$. Sodium bisulfite, calcium bisulfite, magnesium bisulfite, or ammonium bisulfite can be used to form the liquor directly. This process can thus be used in either an acidic, neutral, or basic form.

There is some regeneration of the chemicals used in the sulfite process, but usually not to the extent as with kraft pulping.

Magnesium bisulfite pulping chemicals can be recovered by a Magnefite process, first installed at the Crown Zellerbach mill at Comas, Washington, along the Columbia river.

The primary chemicals used in this pulping process are magnesium hydroxide, $Mg(OH)_2$, and sulfur. The sulfur is oxidized to SO_2 and then reacted with magnesium to form magnesium bisulfite.

The wood chips are digested in this magnesium bisulfite "weak red" cooking liquor for 6 hours and 20 minutes at 325°F, 90 psi. After cooking, the liquor is evaporated from 9% to 55–60% solids, sprayed into a furnace, and burned without additional fuels. The steam generated is recovered for use in the cooking and evaporation steps.

The gas produced contains SO_2 and magnesium oxide, MgO. The MgO is a fine white powder that can be collected by a multicyclone, washed, and reacted with steam to produce $Mg(OH)_2$ for further use. The SO_2 is collected in wet venturi scrubbers and from the stack by absorption into some $Mg(OH)_2$ slurry. This simultaneously generates some weak cooking liquor that can be further strengthened by contact with the gas emitted by the sulfur-burning system.

This system is slightly more expensive, but at the Comas facility, the BOD has been reduced 70%, all but a trace of SO_2 is recycled, the $Mg(OH)_2$ is recycled, and the fossil fuel demand for the cooking and evaporation steps is reduced.

Sodium sulfite or bisulfite chemicals can be recovered by a number of slightly different techniques, all of which are based on reductive burning, as is kraft recovery. The difference is that kraft pulping liquor uses sulfide (S^{2-}), the species of sulfur that is obtained directly from the reductive burning; sulfite or bisulfite processes require instead the SO_3^{2-} or HSO_3^- ion. This results in more complicated recovery processes.

For sodium sulfide recovery systems, the green liquor is bubbled with either pure CO_2 or flue gas (also containing CO_2). This acidifies the green liquor and converts the Na_2S to gaseous H_2S, which is collected and (for most systems) oxidatively burned to form SO_2. The SO_2 can be recycled to the white liquor preparation for reuse.

No techniques for chemical recovery have been developed for calcium sulfite spent liquors due to the formation of insoluble $CaSO_4$. Most of the calcium-based pulping processes are being phased out.

Ammonium base liquors can be burned with or without sulfur recovery, but the ammonia cannot be recovered from a combustion process due to the formation of elemental nitrogen and water.

None of these sulfite recovery systems is generally accepted by the industry, though more are used in Scandinavia than in the United States. Most of the mills dispose of their spent liquor by (oxidative) burning for heat recovery, without chemical recovery, or by waste treatment methods.

Semichemical Pulping

Semichemical pulping methods are a cross between chemical and mechanical pulping techniques, and encompass the entire intermediate range of pulp yields between the two. Generally, wood chips (or logs) are partially softened with chemicals, and the remainder of the pulping is by mechanical forces, most often in disk refiners.

At the high end of the spectrum of yields is the chemi-groundwood process, which is often categorized with the mechanical methods. The whole log is cooked with dilute alkaline sulfite pulping chemicals and then ground. The resulting pulp has greater strength and will drain faster than groundwood pulp, while retaining the advantages: few pulping chemicals, few effluents, and relatively low cost. The method is limited by inadequate liquor penetration.

The most common semichemical pulping method is the neutral sulfite semi-chemical process (NSSC), which uses a sodium sulfite cooking liquor buffered with sodium carbonate to neutralize the organic acids released from the wood during digestion. Other techniques include the green liquor process, which utilizes sodium sulfide and sodium carbonate, and a number of "no sulfur" methods, such as mixtures of sodium carbonate and sodium hydroxide.

In contrast to the sulfite chemical pulping methods, a large majority of the sulfur from the NSSC and related semi-chemical processes can be recycled by a simple acidification/evaporation technique.[4] Once the sulfur is removed, the concentrated spent liquor can be burned for heat recovery.

Other Pulping Processes

A number of new pulping processes have been developed in recent years, largely to minimize the detrimental environmental effects while producing a high-quality pulp. These processes include solvent pulping, ester pulping, and bio-pulping, using various delignifying enzymes produced primarily by white rot fungi. Only the first of these is being used on the commercial scale.[5]

EFFECTS ON THE ENVIRONMENT OF THE PULPING PROCESSES

The specific environmental effects vary, depending upon the pulping process used. The mechanical processes, for example, are quite pollution-free. Most of the potential pollutants involved in the soda process are recovered, making it a fairly clean pulping technique compared to the other processes. There are, of course, still some significant water problems due to the papermaking itself, but these will be considered later. The other two major chemical processes, the sulfite and the kraft processes, generate either air and/or water pollution, which can be directly related to the specific pulping process employed.

Kraft Process

The kraft process generates primarily air pollutants. In this process, many of the chemicals involved in the liquid phase processing are recovered and recycled, so they themselves do not create any serious environmental difficulties. There are some other water problems that do arise, however, and these will be discussed later, with the general water problems associated with all pulping and paper processes.

The release of sulfur-containing gases during kraft digestion and during the burning of the liquor solids (for organic residue removal) do, however, create serious air pollution problems. Typical sulfur compounds that are generated include H_2S, with an odor of rotten eggs, SO_2 and various methyl derivatives such as methyl mercaptan, CH_3SH, dimethyl sulfide, $(CH_3)_2S$, and dimethyl disulfide, CH_3-S-S-CH_3.

These chemicals not only can be a nuisance, but also can be quite toxic. For example, CH_3SH is toxic, and has a worse odor than H_2S. $(CH_3)_2S$ is similar. There is an advantage to the strong odor, though, in that one is usually aware of even very low concentrations of the substance. Table 12-1 lists the typical detection level and various toxicity levels for H_2S.

CH_3SH can be detected as low as 1-2 ppb. Because of the ease of detection, few people are killed by gaseous sulfur compounds. Even though these compounds are, at times, vented from tall stacks at concentrations as high as 2700 mg/m^3 (19,000 ppm). Even after dilution, these chemicals can be detected near the ground, creating a serious nuisance. There is one caution, however: at high concentrations of these gases, one's olfactory senses can become numbed. A person might no longer smell the compounds, even in high concentrations.

There are other environmental effects due to these sulfur compounds, even at low concentrations. For example, they can affect paints, especially those containing lead: $H_2S + Pb^{2+} \rightarrow PbS + 2H^+$. Lead sulfite (PbS) is a black solid whose formation discolors paint.

The sulfur gases can be removed from the gas stream by various techniques. Most companies have installed wet scrubbers. Unfortunately, due to the very low detection limits, these are not always 100% successful at removing the odors.

Various methods can be used for SO_2 recovery. Many of these techniques are similar to those used in power generation, when high-sulfur coal is burned (see Chapter 8).

Sulfite Process

Sulfite paper mills generate primarily water pollutants. Few plants recover any of the processing liquors, so that liquor creates a serious disposal problem. A sulfite process typically produces about 700 lb BOD/ton of pulp produced, compared to the 40 lb BOD/ton of pulp formed by kraft processes.

TABLE 12-1   Detection and Toxicity Levels for H_2S[a]

Substance	Detection	Severe Irritation	Death
H_2S	10 $\mu g/m^3$	30 mg/m^3	60 mg/m^3
	(7 ppb)	(20 ppm)	(400 ppm)

[a]Data from Pyle, James L., *Chemistry and the Technological Backlash*. Englewood Cliffs, New Jersey: Prentice-Hall, 1974.

To date, there is little commercial use for any potentially recoverable sulfite liquor chemicals. Lignin could be a great source of organic chemicals, especially aromatic compounds (it could be used to produce over 60 million tons/year of organic chemicals), but it is unable to compete economically with petro-chemical feedstocks. The sulfite lignin currently can be recovered for several other uses.[6]

1. Lignin chemicals can function as dispersants. Adsorption of the lignin on solids that are suspended in an aqueous solvent causes the particles to gain a negative change, thus to repel each other and remain dispersed. Dispersants are used in products such as pigments, dyes, and agricultural pesticides. They suspend carbonate and phosphate sludges in hard water; they increase the efficiency of the electrolyte in electrolytic refining operations; decrease the viscosity of a slurry; and inhibit crystal growth in aqueous solutions.
2. The lignins also can act as binders in substances such as cattle feeds, ceramics, and refractories. The molecular size and shape of the lignin creates large intermolecular forces between the lignin molecules and those of the particles to be coalesced.
3. Lignins can also be modified for use as sequestrants. Sequestering agents react with metal ions to produce soluble metal complexes, or chelates. In corrosion inhibition, the lignins can, for example, form a zinc complex that acts as a reservoir for slow release of protective zinc ions.
4. The lignins can stabilize oil-in-water emulsions similarly to the way they disperse solids. They can, for example, be used in asphalt whenever the normal asphalt would be difficult to emulsify.
5. Highly sulfonated lignin can be used as humectant, to aid in redispersing dried-out, caked materials, such as that which forms on the inner walls of partially filled containers.

Though these applications are quite diverse, the large quantity of lignin produced daily would soon flood the market. The one large-scale use of sulfite liquor is for binders in drilling mud.

BLEACHING

The nature of the bleaching process, if any, is to a large extent determined by the type of pulping. Kraft pulp, as it exits from the digesters, is a brown color, about the shade of grocery bags. About 55% of kraft pulp is bleached. On the other hand, sulfite pulp is a much lighter, easier-to-bleach pulp. As a consequence, close to 90% of the sulfite pulps are bleached. The mechanical pulps do not undergo a true bleaching process. Instead, they are subject to "brightening." True bleaching removes significant lignin, since chromophoric (color) groups on the lignin are

FIGURE 12-16. Wood pulp that has undergone an extensive bleaching process.

believed primarily responsible for the color. Since the mechanical pulps contain essentially all the lignin that was present in the raw wood, removing lignin at this stage would be counter-productive. Instead, the pulp is decolored by a process that attacks the chromophoric groups, but does not remove them. Unfortunately, none of these methods produces a permanent effect; exposure to light and oxygen causes the lignin to rapidly discolor, as with old newspapers.

Full Bleaching Processes

Bleaching is usually carried out in a sequence of steps, using different chemicals and conditions at each stage. Between each step, the pulp is washed.

There is a shorthand notation commonly used to describe the various stages:

C = chlorine; reaction with elemental Cl_2 in an acidic medium
D = chlorine dioxide; reaction with ClO_2 in an acidic medium
E = alkaline extraction; dissolution of reaction products with NaOH
H = hypochlorite; reaction with NaOCl or $Ca(OCl)_2$ in alkaline solution
P = peroxide; reaction with H_2O_2 or Na_2O_2 in alkaline medium
O = reaction with elemental O_2 at high pressure in alkaline medium
D_C or C_D = admixtures of chlorine and chlorine dioxide

Conventional "full bleaching" is usually conducted in five or six stages, such as CEDED, CEHDED, or OCEDED. Lesser degrees of brightness can be achieved with fewer stages. Sulfite pulp, for example, because of its lighter initial color, could be adequately bleached by, for example, a four-stage process.

The initial bleaching stage, usually using chlorine, does not significantly brighten the pulp; its main purpose is further delignification. Chlorination and extraction are usually conducted in sequence. The chlorination forms chlorinated lignin compounds, which are then solubilized in the subsequent extraction stage. Succeeding stages, though they might remove some further lignin, are primarily designed to brighten the pulp.

Pulp Brightening

Lignin-preserving bleaching treatments are used when a pulp must be decolored without solubilizing the lignin. Significantly different and milder bleaching processes, most commonly employing sodium hydrosulfite or peroxides, are used for this purpose. These chemicals will either reduce or oxidize the chromophores, respectively, but leave the bulk of the lignin unaffected.

ENVIRONMENTAL EFFECTS OF BLEACHING

Toxic Contaminants

Pulp brightening procedures, since they do not solubilize the lignin, have very little effect on the environment. The full bleach processes create the environmental concerns, primarily due to the formation of chlorinated organic compounds in the delignification step. Though sulfite pulp bleaching does generate some toxics, the extensive bleaching that is needed for kraft pulp produces the greatest amount of contaminants.

The contaminants that have caused the most trouble because of their potential toxic effects are the adsorbable organic halogens, or AOX. Of greatest worry is dioxin, or 2, 3, 7, 8-tetrachlorodibenzo-p-dioxin (2378-TCDD), a suspected carcinogen that often is found in trace amounts in finished paper products and wastewater from pulp and paper mills. Dioxin has directly been traced to a condition called chloracne, which is characterized by bumpy rash, blackheads, and pus-filled blisters. Also of concern are phenols, catcechols, and chloroform.

The observed dioxin contamination levels are low. A 1988 survey, conducted jointly by the EPA and the National Council of the Paper Industry for Air and Stream Improvement (NCASI), of all 104 U.S. mills using chlorine to bleach wood pulp, found that median amounts of dioxin were 17 ppt (parts-per-trillion) in sludge, and 24 ppq (parts-per-quadrillion) in wastewater.[7] However, the fear is that dioxins in wastewaters could lead to accumulation in fish, which could then be consumed by humans.

The paper industry has responded to this concern by developing a number of ways to reduce dioxin (and AOX) production. Process modifications have included:

1. Reducing the amount of chlorine used;
2. Replacing some chlorine in the first stage with chlorine dioxide;
3. Adding the chlorine in incremental charges;
4. Using an oxygen delignification first stage;
5. Bleaching with ozone or peroxides;
6. Extended delignification, to remove more lignin within the pulping phases.

In order to decrease the amount of AOX produced, it is necessary to minimize the reaction between elemental chlorine and lignin. Thus, these techniques are designed to reduce either 1) the amount of chlorine used per ton of pulp, particularly in the first bleaching stage, or 2) the amount of lignin remaining in the pulp at the time of contact with Cl_2

It has been estimated[8] that U.S. pulp mills will use 40–50% less chlorine by the mid-1990s. This can be accomplished by the substitution of greater amounts of chlorine dioxide, oxygen, and peroxide during bleaching. Key to the success of this approach is that which occurs in the first (delignification) stage.

Significantly less AOX is produced, for example, if some of the Cl_2 in the first stage is replaced by ClO_2. Unlike elemental chlorine, chlorine dioxide functions by releasing oxygen, not chlorine radicals. Hence, fewer Cl radicals are available to react with the pulp at the stage where the lignin content is by far the greatest. Chlorine dioxide is also a much more effective bleaching agent than is chlorine— nearly three times more effective. Another advantage is that chlorine dioxide oxide can be used with existing bleach equipment.

Recent laboratory studies have indicated that the rates of chemical reaction leading to the formation of dioxins and furans (also considered a toxicity problem) are dependent upon the concentration of the respective reactants (that is, of chlorine and its precursors), rather than the total amount of chlorine applied or the ratio of chlorine to lignin content.[9] The active chlorine can effectively be reduced in a conventional chlorine stage by adding the desired amount in incremental charges, rather than adding the full charge at once.

Hydrogen peroxide, which also can be used in current facilities and with existing equipment, strengthens cellulose fiber as it bleaches. High costs have excluded its use in the past to any major extent, but, since it is a liquid and miscible with water, it is very convenient to use.

The other approach to reduce AOX emissions is to remove the lignin prior to contact with any Cl_2. This can be accomplished by, for example, pulping to a lower yield. The lignin that is solubilized during pulping, particularly kraft pulping, is burned for energy production during the recovery process and is not carried over to the bleaching phase. A limitation of this method, however, is that the pulping must be conducted under chemical and physical conditions that do not also begin to significantly degrade the cellulose.

Delignifying the pulp with oxygen prior to a chlorine stage serves the same purpose: the lignin is no longer present in significant amounts at the time of contact with chlorine. Oxygen alone is not a suitable bleaching method for chemical pulps, however, since O_2 itself only delignifies and does not actually bleach. An oxygen treated furnish is, in fact, quite dark in color.

A number of proprietary chemicals and specialized sequences have been developed in recent years. European companies, particularly in Sweden, Finland, and Germany, have focused on a number of new delignification processes, utilizing both chemicals and enzymes.

End-of-the-pipe treatment methods are also possible. Membrane processes, such as reverse osmosis, can remove up to 90% of the chlorinated organics.

FIGURE 12-17.   The bleaching process, particularly of kraft pulps, can generate an inky-black effluent, even after undergoing secondary waste treatment processing. This photo was taken of the activated sludge process itself at the Finnish Metsa-Botnia pulp mill, and indicates the color of the effluent often emitted to natural waterways.

Color

Though toxic contaminants have been the greatest worry in recent years, another contaminant associated with bleach plant effluent is excessive color. The caustic extraction liquor from kraft pulping operations is particularly black. The dark color is primarily due to tannin. To date, there has been no definitive data linking color emissions to environmental damage or health risks, but it is thought that foam and color can spread over the surfaces of nearby water systems, blocking the sunlight and hence reducing the rate of photosynthesis.

In July 1990, Maine passed the nation's first law regulating color, odor, and foam.[10] Color is measured by visually comparing the effluent to a series of color standards. The restrictions in the law are based on the least stringent of two measurements: the "pounds of color" per ton of pulp manufactured or the change in the color of the water near the mill.

There are a number of techniques that can be used for effluent color removal. The most successful of the early techniques was the use of massive doses of lime to precipitate the dissolved and/or colloidal color bodies.

Recent studies have investigated the use of acidified fly ash for color sorption.[11] On the laboratory scale, upwards of 97–98% of the color can be removed by this technique; there is also some evidence that chlorinated organics are simultaneously removed. The use of one waste to treat a second waste is an intriguing possibility that is currently undergoing additional development.

SECONDARY FIBER UTILIZATION

Secondary fiber is any fibrous material that has undergone a manufacturing process and is being recycled as the raw material for another product. Though the internal recycle of scrap paper, called "broke," could be thus classed as secondary fiber, usually the term is reserved for those fibrous products that have been used in the consumer marketplace (Figure 12-18). In 1988, almost 11% of the pulp used by U.S. manufacturers was from recycled waste paper.

For effective utilization of waste papers, it is necessary to sort and classify the waste paper by grade. Various categories include corrugating, news print, and high-grade deinked. Office papers and similar white papers are of high quality, and demand top prices. Lower grade papers include corrugated boxes and other materials which must be significantly cleaned before reuse to remove hot melt coatings, plastic films, polystyrene film, pressure sensitives, latex, wet strength resins, and the like. Interestingly, laser-printed paper is difficult to deink, hence is considered a lower-grade paper.

Approximately ¾ of all recycled paper in the U.S. is used for corrugating medium and paper board, 10% is used for building products, and 15% is deinked for printing grades. An increasing amount is also used for tissue production.

FIGURE 12-18.   Scrap paper to be recycled. (*Photo courtesy Wisconsin Paper Council.*)

Pulping Process

All recycled paper is first sent to a continuous pulper. This is a large vat with an impellor on the bottom, and frequently a "ragger" and "junker" to remove debris. The ragger consists of several primer wires that are initially rotated in the stock; soon, a debris rope, consisting of strings, wires, and rags, builds on itself and can be gradually pulled from the vat. The junker operates at the side of the vat. Heavy objects are thrown into a recess at the side by centrifugal force, from where they can be removed by a bucket elevator.

Some defibering occurs in the continuous pulper. Many systems also employ a second repulping device, which can enhance the fiber separation and simultaneously remove additional lightweight debris.

The pulp that is destined for printing applications and tissue manufacture must then be deinked. This is essentially a laundering process. Chemicals, heat, and mechanical energy are used to dislodge the ink particles from the fibers and disperse them in an aqueous solution.

The chemicals used are primarily surfactants, or surface-active substances. Specifically, they generally consist of detergents, dispersants, and foaming agents. Other chemicals might include caustic soda, sodium silicate, and borax, which would enhance the surfactant effects.

The pulp then undergoes washing and/or a flotation process, to separate the ink from the stock. A final mild bleaching, perhaps a brightening process, might be used, depending on the ultimate use of the pulp.

Environmental Effects

The effects of pulping waste paper can be separated into two very disparate categories:

1. Effluent emissions; and
2. Impacts on municipal landfills.

The major effluent emission from secondary fiber recovery is polychlorinated biphenyls, or PCBs. PCBs were used in the original carbonless papers. As these papers were recycled, the PCBs made their way into the recycle stream. Though PCBs have been replaced in carbonless papers for many years, the recycle stream is only gradually becoming cleaner. To understand this phenomenon, it is necessary to recognize that given fibers might be recycled many times. For example, if 50% of the paper were recycled and if we start with 100 lb of virgin fiber that is made into paper sheet, 50 lb of this paper would be used a second time to produce a second paper sheet. When that second sheet is recycled, 25 lb of the original fiber would again be used, to produce a third sheet, and so forth. Contaminants such as PCBs are only slowly bled out of the system.

A second factor, making the situation even more difficult, is that some printing inks even now contain PCBs. Thus, these fresh PCBs continue to add to the recycle stream, replenishing some of those that have been removed.

PCBs can, to a large extent, be removed from the effluent by proper operation of a secondary waste treatment facility. The permissible levels are so low, however, that efforts are underway to develop techniques capable of lowering the emission levels to less than 1 ppb.

A complicating factor in a number of locations is that PCBs can build up in the sediment of a river or lake into which the effluent flows. Though the sediment can to a large extent immobilize the PCBs, disturbing the sediment by, for example, dredging remobilizes them into the aqueous environment. In many locations, these "old" PCBs are a much greater source of the contaminant than those present in the current mill effluents.

On the other hand, the overall effect of secondary fiber use on municipal landfills is totally positive. Even if one considers only high-grade printing and writing paper, almost 25 million tons are produced annually. Less than 5 million tons of this are recovered; the remainder is disposed as municipal solid waste.[12]

The common belief is that paper products are readily biodegradable. Though this may normally be true, it is not necessarily so under anaerobic landfill conditions. A study[13] completed in 1989, conducted by boring as deeply as 90 feet into several landfills and taking samples at 10-foot intervals, found that paper and board occupied more than 50% of the volume of waste. The largest single item was newspapers, which occupied some 14% of the volume. Even after burial for as long as 15 years, the newspapers were often readable—and used to confirm the dates of burial.

The United States, in 1988, recycled about 33.3% of its waste paper, or 28.9 million tons. The goal of the U.S. paper industry is to increase that to at least 40%, a slightly smaller percentage than is currently recycled in Japan and the Netherlands. This corresponds to an increase in utilization of approximately 15 million tons of secondary fiber annually, compared to 1988. This cannot have anything but a overall positive effect on future municipal landfills.

There is, however, one caution: recycled paper does produce a sludge. A 1000 t/day deinking mill typically generates > 300 t/day of a ~ 50% ash sludge. Though overall the amount in the landfills is reduced, the waste will be redistributed from large urban centers to papermaking regions.

PULP PROCESSING

After digestion, the pulp must undergo further processing prior to papermaking. The processing steps will vary, depending on the method of pulping and the end use. Screening, thickening, and storage is required for all pulps; cleaning (generally by cyclone cleaners) is required when appearance is important; defibering is necessary for semi-chemical and higher yield chemical pulps; drying is necessary if the pulp is "market pulp" or if it is to be shipped over significant distances to be used at a different mill.

Most pulps undergo "beating" and/or "refining." The two terms are often used interchangeably. Beating is specifically flowing the fiber between perpendicular rotating bars and a stationary bed plate. Refining, on the other hand, is conducted in either conical or disk refiners, where the fibers flow parallel to the bars. Both processes cut and fray the ends of the fiber, to form a stronger, more uniform sheet of paper. Most modern mills have installed refiners, but some of the older mills have retained their old Hollander beaters.

PAPERMAKING

The refined fibers are diluted to at least 99% water and then pumped to a "headbox," which distributes the pulp uniformly across a rapidly moving mesh screen. Constant side-to-side vibration, followed by vacuum pumping and then pressing, extracts most of the water and interlaces the fibers to form a paper sheet. Final water removal is done on dryers, which can either be a series of perhaps 100 smaller cylindrical, steam-heated drums, or one large "Yankee dryer," up to 30 feet across, which operates using heat, pressure, and suction. These steps and related equipment are illustrated in Figures 12-19 and 12-23.

The pulp slurry, greater than 99% water, is fed to a headbox, which distributes the mixture evenly across the paper machine wire (Figure 12-19 and 12-20). The water is withdrawn from the bottom of the wire (Figure 12-21).

As the paper passes down the wire, the water is first removed by gravity and centrifugal action, and then by suction boxes. Figure 12-22 illustrates the original transparent nature of the pulp slurry. As it travels further to the right (as shown), more water is removed by the visible suction boxes. Eventually, the paper is sufficiently dried so that it becomes opaque, and one can observe what is known as the "dry line," on the upper right.

After the wire, the paper is transferred to and further dried by a series of presses and steam-heated rolls. Often, a series of coatings is also applied to develop specific surface properties (Figure 12-23, a and b).

The final paper is then wound onto rolls (Figure 12-24), and otherwise converted to finished products.

Water Pollution

Additional BOD is released to the wastewater in the papermaking. Much of the water removed from the paper sheet carries with it very fine paper fibers and other additives. This contaminated water is frequently called "white water." Reclamation

FIGURE 12-19.   The pulp is evenly spread along the wire. (*Photo courtesy Wisconsin Paper Council.*)

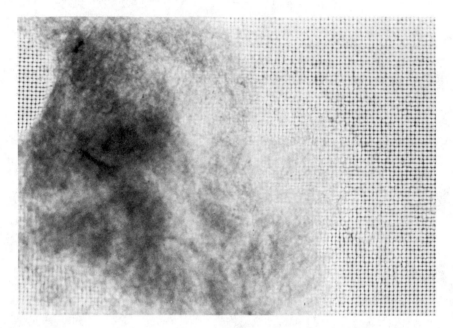

FIGURE 12-20. A closeup of the (99% water) pulp on the wire. (*Photo courtesy Wisconsin Paper Council.*)

FIGURE 12-21. Water being removed from the pulp. (*Photo courtesy Wisconsin Paper Council.*)

FIGURE 12-22. Top view of the paper machine, showing the suction boxes and the dry line, where the paper turns from transparent to opaque.

FIGURE 12-23a. Applying specialized coatings to high-quality paper.

FIGURE 12-23b.

FIGURE 12-24. The paper sheets are wound onto rolls. Each, when completed, weighs several tons. (*Photo courtesy Wisconsin Paper Council.*)

282

of the majority of the white water is economically essential, for it may contain 2–10% of the pulp furnished to the paper machine. Depending upon the production rate and pulp market, that could represent monetary losses in the neighborhood of several hundred dollars/day. The white water can usually be recycled to the beaters (or refiners) for use in dilution, or to the paper machine itself.

If an excessive amount of water is produced for some reason, some must be discharged. Before discharge, the white water must be treated to remove the suspended and dissolved solids. This can be done by combining it with any spent liquor and treating it using traditional techniques.

Overall Water Treatment Processes

At one time these paper mill liquors were just released to nearby streams (Figures 12-25 and 12-26). Now, however, the liquor is either shunted to an oxidizing lagoon, where the organics are decomposed by microorganisms (Figure 12-27), or to a waste treatment facility (Figures 12-28 to 12-29), or it can be concentrated by burning (that is, spent liquor combustion for chemical and/or heat recovery). This latter can, of course, lead to air pollution problems. The spent liquors can be treated either separately or in conjunction with the white water from papermaking. Since the spent liquor is likely to be more contaminated, some mills pretreat that effluent

FIGURE 12-25.   Former effluent from the Badger Paper Mill. (*Photo courtesy Wisconsin DNR.*)

FIGURE 12-26. Aerial view of the effluent from a paper mill diffusing downstream in a major river. (*Photo courtesy Wisconsin DNR.*)

FIGURE 12-27. Aerated oxidizing lagoon for pre-treatment of strong pulp and papermill water pollutants.

FIGURE 12-28. Activated sludge treatment of papermaking effluent. (*Photo courtesy Wisconsin Paper Council.*)

FIGURE 12-29. Settling basins for papermaking effluent. (*Photo courtesy Wisconsin Paper Council.*)

285

in, for example, an aerated lagoon. After several days retention, it can be combined with the white water for secondary treatment.

Sludge

Most mills utilize an activated sludge process for secondary treatment; in some cases, the microbes are aerated with pure oxygen rather than air. In either case, however, pulp and paper mills generate large quantities of waste treatment sludge.

The majority of the mills dispose of this sludge in company-owned landfills (Figure 12-30). Since the late 1980s, however, there has been increasing interest in incinerating the sludge for its heating value. An average sludge might contain 6000 btu/lb and be about 20% ash. Not only can this provide a needed energy source, but it can greatly reduce the amount of landfill space needed. In addition, a number of companies have experienced stability problems in their landfills, due to an excess of water that is disposed along with the sludge. Though it is possible that ash from some particular sludges might contain heavy metals in toxic concentrations, tests to date have indicated that both the fly ash and bottom ash are relatively benign.[14]

Generally, traditional solids processing equipment, often using belt presses for dewatering, can produce a sludge that is only 27–30% solids. About 40% solids is assumed to be the minimum solids content necessary for sustained combustion. Interest has thus focused in recent years on the use, instead, of screw presses, which can achieve 45–50% solids with most sludges. Drying this material further, using

FIGURE 12-30.   Sludge disposal site. (*Photo courtesy Wisconsin DNR.*)

waste heat, is also done at a number of mills. This sludge can be burned in hog fuel burners, as a supplementary or stand-alone fuel, or in fluidized-bed boilers.

Other sludge disposal methods have been investigated, including landspreading, composting, and gasification. Landspreading is a possibility, particularly on papermill-owned forests, but there has been some concern about potential dioxin contamination. The major limitation with composting will probably be the development of sufficient markets, considering the large amount of sludge produced daily. The economics associated with gasification have not yet been verified.

Specialized Treatment Procedures

Several membrane processes—reverse osmosis, electrodialysis, and ultrafiltration—have recently been developed to treat the wastewater that forms at two stages in the pulping process[15]. Electrodialysis procedures have been successfully used to treat bisulfite spent liquor generated in a sulfite pulping process. As in all electrodialysis processes, an imposed electric current is used to cause selective movement of charged ions. A traditional electrodialysis system can be modified for this particular application by placing sulfurous acid (H_2SO_3) in several compartments and the spent liquor (containing large amounts of the lignosulfonic acid sodium salt and smaller amounts of nonionizable sugars) in another set of compartments. The compartments and membranes are arranged such that the sodium and bisulfite ions combine in a third set of compartments, forming sodium bisulfite ($NaHSO_3$). This $NaHSO_3$ can be recycled as cooking liquor. In addition, the remaining mixture of sugars and lignosulfonic acid can be sold as a plywood adhesive.

One difficulty with the process is the high required power costs (up to 500 volts and 80 amps). Hydrogen transfer may also be a problem, due to the lack of charge neutrality within all compartments. More work is currently being done to attempt to improve the process, but the outlook is good.

The greater amount of dilute wastes, arising primarily from the washing of the pulp for further lignin removal, also constitute a serious waste problem. Typically, these wastewaters may contain about 1% dissolved solids. These dilute waters can be treated by a combination of ultra-filtration followed by reverse osmosis, or by just the reverse osmosis step. The purified water can be returned to the system, whereas the small volume, concentrated wastes can then be treated just as the digester wastes.

Environmental Effects of Small Amounts of BOD

A study[16] conducted by the Institute of Paper Science and Technology, has indicated that, contrary to what has been long believed, some of the paper effluents may actually be beneficial to aquatic life. The solids emitted with the effluent are so small (typically 1–2 μm in diameter) that they remain suspended in aquatic systems indefinitely. Some are mineral, but more than half are biological materials, including bacteria and protozoa. About 25% of this biological material, called biosolids, is alive when discharged from a treatment plant. Studies were made with two species of aquatic organisms, the larval aquatic stage of the caddis fly and the water flea of Daphnia. Both species are important, commonly occurring animals widespread in U.S. streams. The results of the tests showed that, with the caddis fly, the smaller, younger members of the sample did quite nicely subsisting on mixed biosolids and normal foods: the larger and more mature caddis fly larvae could survive on biosolids alone. The Daphnia not only survived, but grew and multiplied entirely on biosolids.

These results are interpreted to mean that the presence of biosolids in plant effluents is not detrimental to stream fauna, and that the biosolids might even be a potentially valuable food source for most streams.

If these results are verified, it may be possible that the water pollution regulations for paper mills would not have to be further tightened in the near future, which may save the paper mills millions of dollars in the next few years.

Air Pollution

The papermaking step is extremely energy-intensive. Many paper mills have their own associated power generating facilities. These facilities have the same environmental problems as do other power generating plants, namely particulates and sulfur compounds.

Green Bay Packaging, Inc.[17]

The exact pollution control equipment used at a paper mill will vary from plant to plant with the processes involved and the product desired. The Green Bay Packaging, Inc., mill is an example of a more innovative wastewater treatment system that had been used for over a decade.

The Green Bay Packaging plant is located along the Fox River in Wisconsin. The Fox River shoreline hosts many mills, and it has for years been quite contaminated. In 1966, the mills on the Fox River dumped into the river 438,000 lb of BOD daily, the equivalent of the BOD produced by 2.6 million people. It was estimated, furthermore, that paper manufacture accounted for 82% of the total BOD in Wisconsin's water.[18] In recent years, stringent efforts have been undertaken to clean up both this river and Green Bay (into which it flows).

Green Bay Packaging employed a neutral sulfite semichemical (NSSC) pulping process and a secondary fiber recovery system, manufacturing entirely heavy paper for the manufacture of corrugated paper board and cardboard boxes. In the mid-1970s, they installed a new wastewater treatment system. The following is a brief description of the process:

The liquor from their pulping process, a water solution of the wood pulping reaction products, is separated from the pulp by two screw presses and then screened for reclamation of any usable fiber.

The screened liquor, with about 15% dissolved solids, is split into two streams. The first stream is concentrated by spray-film evaporators to about 21% solids, and then stored for later concentration. The second portion is directly combined with some of the concentrated (21% solids) liquor, and fed to a venturi-type scrubber evaporator. This system allows further concentration to occur, to 43–44% solids.

From here, the 43–44% concentrated liquor is taken to a fluidized-bed reactor (Figures 12-31 and 12-32). This reactor is lined with insulating and refractory brick, and is divided into two compartments, one above the other. The combustion, at 1350–1400°F, occurs in the upper compartment. The air enters from the bottom, and fluidizes a 50-ton bed of particles. The vigorous agitation leads to uniform temperature distribution and adequate combustion air. The reactor is fed by 24 "guns"—23 spraying liquor, 1 spraying sulfur. When the liquor enters the hot, moving bed, it splatters onto the existing particles. The water rapidly evaporates, and the organics burn. The ash, reduced to about 43% of the original volume, deposits on the agitated particles. This ash, due to the added sulfur, is 92% sodium sulfate and 8% carbonate. (Without any added sulfur, the ash would contain only 79% sodium sulfate.) The ash flows downward into the lower compartment, is cooled, and then is collected as fine pellets.

Some of the ash particles can be returned to the reactor to provide a proper volume bed; the remainder of the pellets are sold as makeup chemical to a kraft paper mill.

Green Bay Packaging has also installed a reverse osmosis system for water purification. This system is used to purify any contaminated water from the paper drying process. It is thus an almost totally closed system plant.

This treatment process, which worked very successfully for about a decade, was partially dismantled during the mid-1980s. It had become impossible to obtain the appropriate reverse osmosis membranes from any U.S. vendor. Further modification of the plant was announced in late 1990, when the decision was made to eliminate the NSSC pulping, and focus their efforts on only the pulping of wastepaper.

Scott Paper Company

The settlement between Scott Paper Company and both the EPA and the Wisconsin state environmental control agency is indicative of not only the type of enforcement procedures and fines that are possible—and likely—under current laws, but also of the problems a paper company can encounter when faced with a major effluent cleanup requirement.

The Scott Paper Company plant involved in this settlement is located in Oconto Falls, Wisconsin, on the Oconto River, which flows, like the Fox River, into Green Bay. This pulp mill was built in 1943 and considered marginal in terms of both economics and actual operation. As with many older plants of all types, upgrading

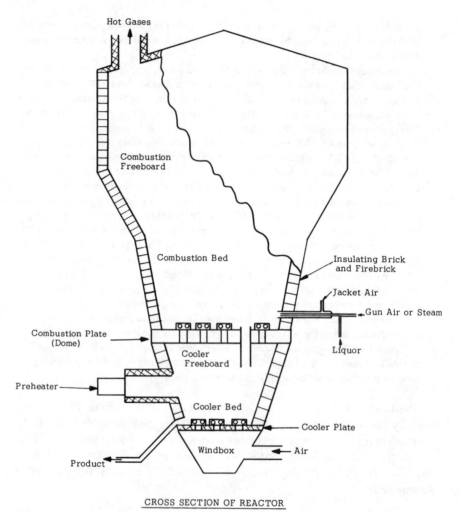

Hot Gases

Combustion
Freeboard

Combustion Bed

Insulating Brick
and Firebrick

Jacket Air

Gun Air or Steam

Combustion Plate
(Dome)

Cooler
Freeboard

Liquor

Preheater

Cooler Bed

Cooler Plate

Windbox

Air

Product

CROSS SECTION OF REACTOR

FIGURE 12-31.   The Green Bay Packaging, Inc., fluidized-bed reactor.

the facility to meet the current stringent environmental regulations was very difficult and costly.

Scott was originally issued a discharge permit in 1974 with the understanding that it would join the City of Oconto Falls in development of a new waste treatment plant. However, three years later, in March of 1977, Scott withdrew from this plan and began construction of its own facility. When the discharge permit expired in July of that year, it was temporarily renewed, allowing a 9600-lb BOD discharge into the Oconto River.

FIGURE 12-32. Aerial view of Green Bay Packaging, Inc. The fluidized-bed reactor is in the center rear, emitting water vapor. (*Photo courtesy Wisconsin DNR.*)

Scott reportedly spent over $3 million to begin construction of a secondary treatment plant, unfortunately without approval of the plan by the EPA or the state. Meanwhile, it repeatedly exceeded the allowed BOD discharge level. For example, the December 1977 discharge average was 98,900 lb/day, 10 times the allowed level and more than 2.6 times the total BOD discharged by the 15 paper mills located on the nearby Fox River.

According to Charles D. Dickey, Chairman and President of the Scott Paper Company: "We were operating in the good faith belief that the standards of an earlier permit still applied. We were in constant communication with government authorities and they were aware that we could not have complied with the new permit except by shutting down the mill." [9]

On February 21, 1978, Scott did close down its pulping operation at Oconto Falls, due to soaring costs. It was no longer a competitive facility due not only to the rising environmental expense (Scott had spent a total of $10 million on environmental improvement for this facility since the mid-1960s), but also to a weakening of the pulp market.

The environmental violations, some 11 charges, that Scott was accused of occurred on selected days between July 1, 1977 and February 21, 1978, when the limits of the pollution discharge permit were violated. Some of the charges, brought by both the EPA and the state, could have led to fines of up to $10,000/day. The federal charges included 10 criminal charges for violation of the Clean Water Act.

Rather than face the lengthy court battle often incurred in these situations, Scott agreed to settle the matter, and entered a plea of no contest, thus paying $1 million in fines. The $1 million was partitioned into several categories: $250,000 was given to the U.S. Treasury and $150,000 to the Wisconsin school fund; $350,000 in federal fines plus $250,000 in state fines were set aside in a trust fund to be used for cleanup operations on the affected water.

Since the closing of the Scott pulp mill, the water quality in the Oconto has improved dramatically. According to James O. McDonald, EPA enforcement director for the Oconto Falls region: "The increase in oxygen and water clarity and the decrease in slimes and bad odors are among the most important outcomes of this case." [20] Stocking of this water by warm-water fish has been very successful.

References

1. *1990 Lockwood—Post's Directory of the Pulp, Paper and Allied Trades*. San Francisco: Miller Freeman Publications, 1990, pp. 1-4.
2. Allan, Leslie *et al, Paper Profits: Pollution in the Pulp and Paper Industry*. Council on Economic Priorities, 1971, p. 1.
3. *Chemistry and the Technological Backlash*. Englewood Cliffs, New Jersey: Prentice Hall, 1974, p. 115.
4. Sell, N. J. and Norman, J. C., "Recovery of Sulfur Dioxide from Spent Sulfite-Based Semichemical Pulping Liquors." Tappi 1991 Engineering Conference, October 1991, Nashville, Tennessee.

5. Fales, Gregg, "Repap Readies Alcell Trial Plant for Further Expansion." *PaperAge* **104**, *No. 10*, October 1988, p. 40.

6. *The Chemistree Book—A Handbook on Lignin Chemicals.* American Can Company, p. 14.

7. Stinson, Stephen, "Paper Mill Wastes Show Low Dioxin Levels." *Chemical and Engineering News* **63**, July 24, 1989, p. 28.

8. Samdani, G., Gilges, K., and Fouhy, K., "Pulp Bleaching: The Race for Safer Methods." *Chemical Engineering* **98**, **No. 1**, January 1991, pp. 37–43.

9. Yann, Arthur, "Westvaco Research Response to Dioxin Challenge." *PaperAge* **106**, *No. 7*, July 1990, pp. 12–13.

10. McCready, Mark, "As Maine Goes, So Goes the Nation?" *PaperAge* **106**, *No. 7*, July 1990, p. 8.

11. Nasr, M. S., Gilles, R. G., Bakhshi, N. N., MacDonald, D. G., "Color Remover of Pulp Mill Effluents with Fly Ash." Proceedings of the Second National Conference on Complete WateRuse: Waters Interface with Energy, Air and Solids, May 4–8, 1975, Chicago, Illinois.

12. Jannazzi, F. D., and Strauss R., "Waste Printing and Writing Paper: Problems and Opportunities for the Paper Industry." *PaperAge Recycling Annual 1990*, pp. 42–43.

13. Reese, K. M., "Archaeologists Dig Into the Secrets of Landfills." *Chemical and Engineering News* **63**, July 3, 1989, p. 60.

14. Sell, N. J., McIntosh, T. H., Jayne, T. G., Rehfeldt, T., and Doshi, M. R. "Burning Bulk Screw Press Dewatered and Briquetted Sludge in a Hog Fuel Boiler." *Tappi J.* **73**, *No. 11*, November 1990, pp. 181–188.

15. "Membrane Processes Upgrade Wastes." *Environmental Science and Technology* **3**, *No. 11*, November 1969, p. 1147.

16. "Paper Mill Wastes May Benefit Aquatic Life." *Chemical and Engineering News*, June 19, 1978, p. 23.

17. *Green Bay Packaging—Recovery Plant Operation.* Green Bay Packaging, Inc.

18. Allan. p. 9.

19. Otto, Dave, "Scott Paper Fined $1 Million." Green Bay Press-Gazette LXIII, *No. 188*, p. A-1.

20. *Ibid.* p. A-1.

13

Food Process Industries

Today's eating habits, with the emphasis on precooked, packaged convenience foods, have led to a very rapid rise in the per capita consumption of commercially prepared foods. Along with this increased consumption comes also an increase in the pollutants, particularly water pollution. The American food processing industry annually produces some 800,000,000 lb BOD, 392,000,000 lb soluble solids, and 800,000,000 tons of solid waste residuals in canning and freezing plants alone.[1] Additional wastes are produced by the meat and poultry slaughtering, fish processing, and dairy industries.

The food wastes are those portions of the raw materials for which there is little demand. Many of the peelings, for example, that are not being used for human consumption or converted to pet or livestock feed are released as BOD (Figure 13-1).

The food processing industry uses large amounts of water. In a typical year, 83,000,000,000 gallons of wastewater are produced in treating 26,000,000 tons of raw products.[2] The water supply and waste disposal often limit the growth of the plant, and determine the resulting cost of the product. Water saving techniques are becoming increasingly important as production increases, and can sometimes mean a significant savings in total production expense.

As the industries have expanded, the regulatory authorities have tightened the controls on the effluents. Many municipalities have expanded their water treatment facilities and now treat the water from food processing industries, for a service charge, along with the remainder of the municipal wastes. Many plants, however, have found it profitable to at least pretreat their wastes. One reason for this is the seasonal character of many wastes. Frequently, it is economically feasible to separate the various types of wastes and treat them as such with on-site facilities. The sludge generated by these treatments is particularly difficult to handle, though, because of its high moisture and fiber content, and its tendency to rapidly decompose.

FIGURE 13-1.   Stokely Van Camp waste beets that must be disposed of. (*Photo courtesy Wisconsin DNR.*)

Regardless of the treatment chosen, minimization of the water used (up to a 40% savings is possible)[3] and of the wastes generated is very advantageous. The wastes that are formed can, preferably, be converted to useful byproducts. Many of the sludges contain large quantities of protein, the B vitamins (B_{12}, thiamin, riboflavin, and niacin), pectin, and other nutrients that can be recovered for use or used directly as a food supplement.

CANNING

Much progress has been made in recent years in the development of new methods for treating canning wastes. Though each cannery is probably unique in the specific production processes employed, the processes involved can be categorized as: 1) preliminary cleaning and preparation, 2) canning of fruits and vegetables, and 3) canning of juices.[4]

The first step is always to clean the raw product of insects, dried juices, chemical residues, and adhering soil. The cleaning method employed varies with the physical characteristics of the food, but typically involves water baths and/or high-pressure water sprays, with the materials moving on conveyor belts or passed through revolving screens. The water used may either be fresh, or good-quality water recycled from some other in-plant operation. The used water will contain both dissolved and suspended contaminants. The quantity varies greatly with the harvesting methods: the mechanically picked products include more debris, leaves, and vines, as well as more damaged product, than do those raw products that are handpicked, but the economic advantages of mechanical picking far outweigh any other considerations.

After the preliminary separations and cleaning procedures, the raw product must be further cleaned, inspected, and sorted. The methods are diverse, and include both wet and dry procedures and mechanical and hand operations. For example, tomatoes are typically moved through a soaking tank and then washed by an overhead spray. Shelled peas, on the other hand, are first placed on shaker screens to remove pieces of pod, vine, and fine dirt. An air blast removes the very light materials. Then the peas are taken to a flotation washer, which employs cold water to dissolve the juices; the heavy sand, small stones, and silt sink, while the skins and other lighter-weight impurities float. This step can be even more efficient if a froth flotation cleaner is used. If there is a difference between the wettability of the product and the impurities, such as between peas and nightshade berries and various weed seeds, the product that more easily wets will sink (in this case, the peas), while other substances will float in the froth emulsion. Finally, the peas are separated out from the washwater by a reel-type device equipped with a fresh water spray.

The products must then be sorted for size, maturity, weight, quality, or similar characteristics. Again, this can be done manually or automatically.

Next, the materials must be trimmed, cored, pitted, and/or cut. Substances such as pears, apples, peaches, olives, cherries, and apricots are usually pitted before further processing. Most are halved and pitted mechanically and then hand-inspected.

Peeling is required for many fruits and vegetables. Mechanical or chemical peeling is employed for most. Some types of raw products can be peeled by revolving them in a cylinder with abrasive walls. Many others, particularly tomatoes, peaches, and potatoes, employ a hot lye solution for peel removal. (This method is further discussed later.)

After peeling, the product must often undergo some final preparation prior to the actual canning or juicing. Carrots and potatoes often are diced; other fruits or vegetables are sliced. Often, several types or varieties of product are blended—and all must undergo final inspection.

The washed and prepared raw products are transported between various operations by belts and pumping systems. Large quantities of pure water are required to do so. Typically, 50% of the total plant water is consumed in these preliminary operations.

The actual canning of the fruits and vegetables also involves many possible steps. Most products must be blanched, or exposed to high temperatures for a short period of time, to retard bacterial and enzyme action that would cause rapid degeneration of the product quality. Baby food, as well as some other products, must be pulped and strained either prior to or after cooking and heating. Tomato sauce, catsup, and tomato paste also require cooking. Spices, thickening agents, syrups, juices, water, salt, and other ingredients may also be mixed with the various products.

Packaging can be done in metal cans, glass containers, lined 55-gallon steel drums, and plastic-lined fiber-backed drums. These must be cleaned prior to filling and after sealing. This water can generally be recycled with little difficulty. Lastly, the sealed cans are subjected to thermoprocessing for varying lengths of time. The purpose is to sterilize the product by killing all possible bacteria that might harm the product and/or the consumer.

The production of fruit or vegetable juices requires several additional steps. After the initial preparation, the juices must be extracted from the remainder of the product. Many methods are available for this process. Citrus juices, for example, can be extracted by reaming the halved fruits, or by pressing the fruits by various types of rotating rolls. Tomato juice often is extracted by compression of the tomatoes between a screw and screen, or by a paddle-type extractor.

Regardless of the type of extraction, it is then necessary to screen the juice to remove the pulp and seeds. Citrus juices must additionally be "de-oiled" by heating to approximately 125°F under reduced pressure. Eighty percent of the oils are thus removed by vacuum distillation. This de-oiling also de-aerates the juice. If de-oiling is not done, de-aeration must be done separately. A thin film of juice is run through a vacuum chamber. Contaminated steam is produced, and it is sometimes vented to the atmosphere, or is condensed and treated for pollutant removal.

Before concentration, the juice should be pasteurized by heating to 185–200°F for a few seconds. This has the same effect as blanching. The juice can then be concentrated in a "vacuum evaporator," basically a vacuum distillation process. The use of a partial vacuum allows the concentration to occur at moderate temperatures, in order to not change or decompose the product. Some of the volatile constituents might need to be added back to the concentrate to retain good flavor. It is possible to fractionate the distillate and return the desired components, or to add the desired components from some other source.

Juices such as tomato juice often are homogenized to prevent settling of the solids. In order to do so, the juice is heated to about 150°F under 1000–1500 psi of pressure. This produces a thicker product, which is more desirable to the customer.

The juice cans are then sealed, as are the fruit and vegetable cans.

Two phases in the processing—peeling and blanching—appear to be responsible for the majority of the canning pollution.[5]

Peeling

Peeling must be done for many fruits and vegetables, notably potatoes, tomatoes, and peaches. More potatoes are produced in this country than any other vegetable. It is not unusual for a canning plant to process 1,000,000 lb potatoes/day, producing a BOD equivalent to that produced by a city of about 300,000 people. The majority (75%) of this BOD comes from the peeling step.[6]

The conventional method of potato peeling involves dipping the potatoes for several minutes in a 16-20% lye (NaOH) solution held at 95-120°F, draining them and then maintaining them for 2-5 minutes near the boiling point. The NaOH softens the skin and the portion of the potato immediately under it. After this treatment, the potatoes are peeled by high-pressure water jets, giving rise to a highly alkaline effluent. This method requires 2 1/2-5 lb water/1 lb potatoes, 5-10 lb NaOH/1000 lb potatoes, and results in a loss of 8-20% of the potato, the "peel loss." [7]

A new process, dry caustic peeling, has been developed at the USDA Western Regional Laboratory, to reduce pollution and also produce a better product. The potatoes can now be dipped for only 1 minute in a 12% NaOH solution, held for 3-5 minutes, and transported on a roller conveyor through a gas-fired infrared heating unit for a 1-minute heating period. The infrared heating dries the NaOH solution, which speeds up the peel-softening process. The potatoes are then mechanically peeled by a "Magnascrubber," a tumbling unit with rotating rolls equipped with 1/2-in rubber studs. The peel, after removal by these rubber-tipped rolls, accumulates as a 25% solids residue, about the same as found in an average potato. The potatoes are washed in a finisher that employs wire brushes with a water spray to remove the cooked, rather gelatinous materials, and are then put into a holding tank to prevent oxidation.

This dry caustic process requires less than 1 lb water/lb potatoes, 2-3 lb NaOH/1000 lb potatoes, and results in a loss of only 5-12% of the potatoes. In addition, the wastes generated are much easier to handle. If treated by conventional means, the effluent is only 0.7-1.9% solids; that from the newly developed peeler is as high as 25% solids, and, even when combined with the finished wastes, the overall solids content is typically about 15%. With the dry caustic process, 85% of the total peel waste solids can be kept separate from the remainder of the plant effluents. Originally, the waste has a pH of about 12 (fairly alkaline), but the peel waste spontaneously self-ferments down to a pH of about 5. At this stage, it can be mixed with the trimming wastes (potato eyes, and spoiled, dropped, or defective potatoes) and used for animal feed. The wastes from a conventional treatment process, on the other hand, are difficult to treat. Primary treatment typically removes only about 50% of the solids: the rest must be removed by some secondary treatment, discharged into lagoons, or used for land irrigation.

This peeling technique is not restricted to potatoes. Cling peaches can be similarly peeled, using, instead of the rubber studs, rubber discs of various dimensions. The water use can thus be reduced 90%, and the peel—12-13% solids—forms a pumpable slurry for land disposal or animal feed. Other uses of this peeling technique include tomatoes, beets, and carrots.

Blanching

Blanching is an essential step in the processing of many vegetables. It consists of heating the product to destroy the enzymes that would lead to bad flavors, color changes, and/or tissue softening. Conventional blanching has been done in hot water or steam. Generally, the vegetables either pass on a conveyor belt through a high-temperature box, or are dipped in water, in order to raise their temperature to 155°F. The water or steam is an inexpensive heat-exchange medium, but the resultant leaching of various substances has led to serious waste problems due to the high BOD. In fact, blanching is now the second most polluting step in the canning process.

Two techniques have been developed to minimize these environmental effects: hot gas blanching and the individual quick blanch.

Hot gas blanching employs the combusted gases from a natural gas furnace. These gases, mixed with a little steam to prevent dehydration of the product and increase the heat transfer characteristics, are blown through the vegetables, which are spread on a conveyor. The immediate result is an almost fluidized bed of that vegetable. Vegetables so blanched—spinach, beets, green beans, corn on the cob, peas, asparagus, and pumpkin—taste perhaps slightly different from the conventional, but the flavor reportedly is good. Moreover, they may be more nutritious; spinach so processed retains more ascorbic acid and more of the water-soluble vitamins.[8] In addition, as hoped, the technique drastically cuts the pollution generated. The wastewater volume is typically reduced 99% and the BOD reduced 96% per ton of product.[9] The process is more expensive for most vegetables (except green beans) than is the conventional process, but it is hoped that this is more than offset by the savings gained by not having to treat and dispose of as much waste.

The individual quick blanch is another possibility that has been tested. Foods such as lima beans or peas are first partially dehydrated (5-8%), and then spread in a single layer and exposed to steam just long enough to raise their average temperature to that required for blanching. Next, the vegetables are quickly gathered into a very thick bed. This bed can retain the heat long enough for all of the constituent vegetables to achieve uniform blanching. The initial partial dehydration allows the product to absorb the steam as it condenses, essentially eliminating all of the losses due to leaching.

Water Treatment

Overall, the canning processes recycle as much water as possible. In spite of this, all of the water does become contaminated and must eventually be treated. Most of the treatment methods employed are similar to those used in municipal facilities. Trickling filters, employing a synthetic filter media of honeycombed polyvinyl chloride, can be quite successful as a form of biological treatment. Settling lagoons are also possible, as are activated sludge procedures (Figure 13-2). Usually, some additional treatment is required, however. This treatment can be chemical, such as the addition of acid ferrous sulfate and lime to the screened wastes,[10] or it can simply be a two-stage primary and secondary process such as that instituted at Stilwell, Oklahoma (below).[11]

The Stilwell Canning Company

The Stilwell Canning Company, of Stilwell, Oklahoma, cans and freezes a large variety of fruits and vegetables, including strawberries, spinach, okra, peas, beans, and white and sweet potatoes, with potatoes dominating. Previous to 1969, the Stilwell wastes were split into two effluent streams. The stronger wastes from the potatoes were pumped three-fourths of a mile over a 500-ft hill into a holding

FIGURE 13-2.   Foremost Foods aeration tank for secondary wastewater treatment. (*Photo courtesy Wisconsin DNR.*)

lagoon, for later land disposal on a fruit orchard. The system was poorly operated, resulting in several problems, the most noticeable being an intense odor spreading downwind several miles. To control these odors, sodium nitrate was dumped into the lagoon. Unfortunately, this sodium nitrate killed off some of the fruit trees, and consequently, the orchard owner refused to allow the plant to irrigate his land with these wastes.

The lower-BOD wastes, those from the other vegetable products, were discharged into the Stilwell city sewage treatment system, designed to handle the wastes from 6000 people. But the company waste varied between the equivalent waste produced by 7200–15,000 people, and this, coupled with the Stilwell population at that time of 2600, completely overwhelmed the facility.

The effluent from the municipal facility was discharged into Caney Creek, a spring-fed stream. When the cannery was in full operation, no dissolved oxygen was found until 5 miles downstream. Caney Creek eventually—after 16 miles—flows into Lake Tenkiller, a major recreation area. Numerous complaints were lodged against the operation, due to the resulting bad odors. So the company clearly needed a treatment facility capable of handling the high-strength, nutritionally unbalanced, large-volume wastes.

Designing an adequate and economically and technologically feasible facility was difficult. Measurements showed that the organic load varied by as much as 60 times. A volume of 0.39 million gallons/day with a COD of 150 mg/L for spinach wastes was contrasted with a 1.91 million gallons/day volume with a COD of 5500 mg/L while processing sweet potatoes.

A number of treatment systems were considered before staff engineers decided upon the current two-stage process. The first stage employs primary clarification and minimal solids aeration. This stage is characterized by a high loading rate and a short solids retention time. Although its removal rates are high, its removal efficiency is low. Therefore, the second stage must have a large removal efficiency at a low rate. More aerators, allowing activated sludge digestion, thus comprise the second stage.

Operation of the treatment facility consists of initial gross solids removal by screens and subsequent trucking of these solids to a nearby landfill. The primary clarifier and vacuum filter solids are also disposed of by landfill. The liquid waste flows from the screens to the minimal aeration basin designed to remove 50% of the COD and then to two large aeration basins connected in parallel, the aerobic sludge digestion units. Final clarification is then carried out to remove any remaining solids.

This system provides for much flexibility in operation. When vegetable processing is underway, for example, only the extended aeration units are needed. When potatoes are processed, both stages are put into operation. Because of this flexibility and for other reasons, the plant has a greater than 95% removal efficiency.

PICKLING

Pickling is a process employed by many food industries. Some produce exclusively (cucumber) pickles. Others pickle olives, tomatoes, or sauerkraut.

There are various techniques available that can be used to accomplish the pickling. Traditionally, the vegetables, often small cucumbers or gherkins, can be soaked in a brine (NaCl) and vinegar solution and fermented. Seasonings such as sugar, mustard, dill, horseradish, cinnamon, allspice, cloves, celery seed, peppercorn, and pimento may be added. After this fermentation period, the pickles are put into jars and the jars sealed.

About 40% of cucumber pickles are now made directly into "fresh-pack" products. This method avoids the customary brining. Fresh cucumbers are packed into jars with brine, vinegar, herbs such as dill, and some essential oils. These cucumbers are then pasteurized in the sealed jars. The result is a fresher texture and color than possible with the normal brining procedure; a restriction, however, is that the cucumbers must be processed very soon after harvesting.

Another available processing technique is "pure culture" fermentation. Pasteurized cucumbers, acid-forming lactic acid bacteria, salt, and other flavorings are put directly into the retail jar. Inside the jar, the bacteria will react, producing the desired acidity and flavor.

Little recovery work has been done on the brines. The latter process, pure culture fermentation, produces no wastes, and fresh-pack produces only a little more. The traditional brining process produces large amounts of brine, which are usually wasted.

Chemicals Recovery*

One associated recovery process has recently been developed. The standard method of testing for salt content in pickles, and in many other types of food, is to grind up a small portion of the product, and to titrate the resultant solution with silver nitrate ($AgNO_3$): $AgNO_3$ (aq) + NaCl (aq) → AgCl(s) + NO_3^-(aq) + Na^+ (aq). The endpoint could be determined by observing when the silver chloride (AgCl), a white solid, ceases to form, but better results are obtained by the use of a potassium chromate (K_2CrO_4) indicator. Until the endpoint, only white AgCl forms. After the precipitation of AgCl is complete, orange-red silver chromate (Ag_2CrO_4) precipitates: CrO_4^{2-} + 2 Ag^+ → Ag_2CrO_4(s). The AgCl is less soluble than the Ag_2CrO_4, so the latter cannot form permanently in the mixture until precipitation of the chloride ion (Cl^-) as AgCl has lowered the Cl^- concentration to a very small value.

The resultant solution, after titration, thus consists primarily of AgCl, Ag_2CrO_4, sodium ions (Na^+), nitrate ions (NO_3^-), water, and a large number of various organic

---

*Information provided by David W. Osten, Osten Chemicals Research, Inc.

substances. Considering the present cost of silver, and the EPA restriction that specifies that these wastes must be properly disposed, it is very feasible to recover this silver. The silver metal can then be sold, or can easily be reprocessed to $AgNO_3$ at the food processing plant, to be reused as new titrant.

The relatively high concentration of AgCl and the presence of the organics prohibit use of the traditional recovery processes (such as those employed in the photographic industry). Economical recovery can, however, be accomplished by taking advantage of the fact that silver is not a particularly "active" metal. Its standard oxidation potential ($Ag \rightarrow Ag^+ + e^-$) is -0.7991 volts compared to -0.337 volts for copper ($Cu \rightarrow Cu^{2+} + 2 e^-$). Copper thus has a greater tendency to ionize, and can replace any agent in solutions: $Ag^+ + Cu \rightleftharpoons 2 Ag + Cu^{2+}$. To keep the Ag^+ in solution, it is necessary to add ammonium hydroxide (NH_4OH). The result is a solution of silver ammonium ions, $Ag(NH_3)_2^+$. A copper electrode placed in this solution quickly gathers a large amount of impure silver metal; this impure silver does not adhere to the copper, and thus can be quickly shaken off and collected by decanting, filtering, and/or centrifuging the mixture. The solid is then heated to first drive off any contaminating organics, and then taken to above the melting point of silver ($960°C$). At this temperature, any remaining contaminants such as copper will form a scum and can be separated from the pure metal. If desired, further purification can be accomplished by electrolysis. This pure silver can then be converted to $AgNO_3$ by simply dissolving it in concentrated nitric acid (HNO_3).

This procedure results in a silver recovery greater than 80%, particularly if the waste does not contain excessive solids. Even a small pickling plant spends about $8000/year for $AgNO_3$. In addition, there are the disposal costs of the titrating solution. If 80% can be recovered, the savings would be about $6300/year on chemicals alone. The net expenses for reagents and maintenance are less than $1000, and the capital investment about $500, so the savings are obvious.

MEAT PACKING[12]

There are three types of plants that can be classified as meat packing facilities: slaughterhouses (killing and dressing plants with no associated processing of byproducts), meat processing plants (which do no slaughtering), and meat packing plants (which both slaughter and process the meat). Typically, meat packing plants will cook, cure, smoke, and pickle meat; manufacture sausage; and render fats into lard, tallow, or greases.

In the meat packing industry, the cattle, calves, sheep, and hogs are first detained for a few hours and then immobilized by chemical, electrical, or mechanical means. The cattle are generally suspended from an overhead rail, and the hogs are positioned on a conveyor with their heads over a trough, for sticking and bleeding.

Blood recovery is crucial, for blood is one of the major sources of BOD. Failure to recover the blood reportedly increases BOD by 72%.[13] It is estimated that the blood from a single animal, if totally discharged to a sewer, would create as much BOD as that created by 50 people. The majority of meat packers, particularly the larger plants, do currently recover the blood.

The next step is hide removal, or, in the case of hogs, de-hairing without skinning. Hide removal is usually done with mechanical hair removers, at a rate of 50-60 heads/hour. Hog hair is removed by first loosening the hair by soaking the carcass in water at about 140°F, then scraping off the hair. The hair can be sold (often to manufacture foam rubber), disposed of by landfill, or dissolved in a strongly basic solution. The viscera are then removed, and separated depending upon whether or not they are edible. The hearts, stomach, and lower digestive tracts usually are used as food.

All ruminants, such as cattle and sheep, have two stomachs. The first, called the paunch, contains large amounts (60-90 lb) of undigested materials (Figure 13-3). This material, typically hay, corn, straw, and grain, must be removed and disposed of. The disposal methods fall into three classes: 1) dry dumping, 2) wet dumping, and 3) dumping into the sewer system.

Dumping of the materials into the sewer is the oldest practice, and was once practiced primarily by those packers located along major rivers, where the effluent could be released with minimal treatment. Paunch material is very difficult to treat,

FIGURE 13-3.   Paunch wastes for disposal. (*Photo courtesy Wisconsin DNR.*)

particularly by a biological treatment facility, and it is far preferable if the materials are handled separately from the beginning.

Most plants today practice wet dumping of the paunch contents. Pressurized water is used to flush the paunch contents into a stream of water, which is then passed over vibrating or rotating screens. These screens separate the materials according to size. The coarse materials are trucked away for landfill disposal, and the fines and any dissolved materials flow into the sewer. This gross separation removes up to 95% of the possible BOD content.

Dry dumping of these materials is the preferable disposal method. The undigested materials are conveyed mechanically to a truck for use as landfill or soil conditioner. The paunch itself is then rinsed and either used for tripe or for inedible rendering. There has been little success in processing the paunch materials into animal feed.

Rendering of the fats, to either edible lard and tallow or to the inedible fats used in soap and grease manufacture, is very important in waste recovery (Figure 13-4). Edible rendering is applied primarily to hog and some beef fat; inedible rendering is usually applied to scraps, trimmings, and inedible organs such as heads and feet of various animals. Those parts of the animal that have been condemned by a meat inspector are also often rendered to an inedible product.

There are several possible rendering methods that can be employed. Edible and inedible rendering are both usually done by one of the following: dry rendering, wet rendering, or low temperature rendering.

FIGURE 13-4.   Wastes to be rendered for soap. (*Photo courtesy Wisconsin DNR.*)

Dry rendering is accomplished by cooking the fats in steam jacket tanks under a vacuum. The vapors from this process, if vented without washing, may cause unpleasant odors. There are essentially no water pollution problems, and almost total recovery of the protein materials is possible.

Wet rendering involves loading a large tank with fat, sealing it, and then introducing steam at 40–60 lb of pressure. The result, upon standing, is the formation of an upper fat layer, a middle water layer, and a bottom slush layer. The fats must be drawn off, and the water and slush treated by a centrifuge or some type of press to remove the suspended solids. The liquid, after solids removal, contains about 75% of the proteins, and thus has a typical BOD content of about 32,000 ppm.

It is usually economically favorable to concentrate this liquid, the "tank water." The concentrated materials, if from an edible rendering process, can be sold as raw materials for pharmaceutical houses. In spite of the concentration and recovery of these materials, this process still creates a high BOD flow to the sewer, and thus is not a desirable treatment method in these days of pollution awareness.

Low temperature rendering involves the mechanical breakdown of fatty tissues at temperatures no greater than 118°F. This is essentially a dry process, for no water is added, so the water pollution generated is very minimal. The lower temperatures should also decrease the vapors, and thus the odors. Currently, this is the most advanced technology available for rendering.

The remainder of the carcass must, of course, be further trimmed, cut, possibly deboned, and possibly processed by grinding, curing, pickling, smoking, cooking, and/or canning.

Pollution Control

Overall, the meat packing wastes are either solids or liquids. The solids, comprised of manure from the livestock pens and the paunch contents, are usually composted or landfilled. The wastes from the remainder of the steps are converted to liquid wastes. Generally, these wastes first go to a grease trap or flotation unit and then to some form of secondary treatment.

Many plants take advantage of the local municipal facilities for treatment of their wastes. Some plants have pretreatment facilities; others do not. The desirability of pretreatment depends upon many factors, including the city's rate structure, the market for the grease and solids that can be recovered, and the possible complaints from the residents and the municipality if pretreatment is not done.

Pretreating of wastes is done by either screening or centrifuging. However, since many of the solids (the meat particles, fat, and manure) break down readily when subjected to turbulence and pumping, releasing soluble BOD and suspended and colloidal grease solids, it is far preferable if these solids can just be kept out of the sewer from the beginning. Much can be separated out for landfill or soil conditioning if a plant is careful.

As with all industries, it is desirable to recover as many solids and byproducts as possible. Water conservation is also very important, as it is in the canning industry. Incautious use of water is not only a waste in itself, and increases the volume of liquid that must be treated, but it can also lead to the washing into the sewers of many solids that could be separated and otherwise treated.

If the plant dumps its wastes in the sewers, the treatment method is determined by that chosen by the municipality. If the plant decides to treat its own wastes, however, some of the preferred secondary treatment methods include: 1) anaerobic processes, 2) aerobic lagoon systems, 3) activated sludge processes, 4) trickling filters, and 5) rotating biological discs. All of these methods have been discussed previously except the last, the rotating biological discs.

The rotating biological disc system (Figure 13-5) is comprised of large-diameter, lightweight plastic or high-density styrofoam discs mounted on a shaft. This shaft is placed horizontally in a semicircular tank through which the waste water flows. As the discs rotate, they aerate the water. Many of the organisms present in the wastewater can adhere to the discs and begin to multiply. Some of the wastewater also adheres to the discs. This wastewater simultaneously is contacted by aerobic microorganisms and oxygen. The result is growth of the organisms and simultaneous consumption of the suspended and dissolved organic materials.

FIGURE 13-5.   Rotating biological discs that are used for secondary treatment of, particularly, food process wastewater. (*Photo courtesy Autotrol.*)

Air pollution problems must, of course, be solved at the site. Many complaints from local residents center on the odors created by the rendering (and perhaps the smokehouse) operations. One of the most objectionable types of odors is that of decaying meat. This aided our ancestors in determining if meat was spoiled. The human nose is capable of detecting some rendering odors, mainly amines and mercaptans, to a few parts per thousand.

None of the possible odor removal methods yield any useful byproduct. Thus, in order to be attractive for a plant, the method must be inexpensive as well as efficient. These odors really can be a serious problem. Without odor controls, the rendering odors have been noticeable at distances up to 20 miles away, and have led to many complaints by citizens.[14]

In accordance with the original clean air act of 1970, standards limiting the emission of undiluted odors could be set. These standards would be in terms of "odor units." As discussed in Chapter 1, an odor unit is defined as the quantity of a single substance, or combination of substances that, when completely dispersed in 1 ft^3 of air, is detectable by the median number of observers in a panel of eight or more. Because of this type of measurement, and the fact that the strength of the odor is so intrinsically related to the particular individuals on the panel, the whole procedure is quite subjective. However, to get an idea of the magnitude of the "odor units," roasting coffee or fresh baked bread emit about 2000 odor units/ft^3.[15] The EPA has not to date set any regulations as to odor limits, though 200 odor units/ft^3 have been discussed. Some states do have regulations; for example, the Wisconsin Department of Natural Resources takes action if 60% of a random sample of the population feel an odor is objectionable. Maine now has a law regulating color, odor, and foam.

There are three methods commonly used for odor control: combustion (including open flare burning, direct flame combustion, and catalytic oxidation), scrubbing, and chemical absorption and adsorption. Combustion, or thermal incineration, appears to be the most effective method for removing strong rendering odors from the effluent air. If the odoriferous substances constitute a potential fuel source, the economics of the process are enhanced. Care must be taken to assure complete combustion of the various substances, for partial oxidation products have the potential of being even more odoriferous than the original substances.

Catalytic oxidation is another combustion possibility. It can occur at lower temperatures, hence with a lower fuel cost, but is not economically capable of treating large volumes of air. In addition, the capital costs are high. Perhaps the major problem, however, is the rapid catalyst poisoning that typically occurs, forcing the systems to be operated at temperatures greater than 1200°F.

Water scrubbing of the odors has also not been particularly successful because many of the components that produce the odors are insoluble in water. Reactive scrubbing is possible, and requires that inert precipitates or easily removable nonodorous complexes be formed. Possibilities include water solutions of sodium bicarbonate, sodium bisulfite, sodium hydroxide, chlorine dioxide, chlorine, potassium permanganate, and calcium hypochlorite. Adsorption with activated carbons can be successfully done, but it requires constant monitoring to avoid breakthrough of odorous substances. Other problems include variation in adsorptive capacity, low capacity, and, if not regenerated, relatively short life of the absorbent.

Other methods have also been studied. Ozonation of the rendering odors, for example, has been successful, particularly in Japan. Masking of these odors is not sufficient in itself.[16]

Packerland Packing Company

Packerland Packing Company, located in Green Bay, Wisconsin, is typical of many packing plants. It produces a very high BOD wastewater, and simultaneously has the potential to generate very objectionable odors.

Since the early 1970s, a number of different pollution control systems have been implemented. The following is a brief explanation of the Packerland facilities and a short description of the evolution of the pollution control systems to the present methodology.

Packerland Packing Company uses about 350 gallons of water per head of cattle slaughtered. The average slaughtering rate is 2300 head per day, 5 days per week, resulting in an average flow of about 800,000 gallons per day. This water is used in two main areas: about 2/3 is used for slaughtering and meat processing, and 1/3 for rendering operations. The water has, particularly, a high blood content from the kill floor and a lot of fat from the casing and pet food departments.

At the kill floor, the water initially flows through sizable static or circular roller screens that remove the larger solids (horns, meat, etc.). The water then goes through a skimming tank, which removes a top skim (mainly fats) and a bottom sludge.

The fat, along with the bottom sludge, the offal (heads, entrails, horns, and similar waste parts), and the blood, is sent to the rendering plant. At the rendering plant, the raw inedible materials are brought to a temperature of 245°F to sterilize them and to evaporate much of the water content. The resulting material is about 50% water, 20% fat, and 30% solids. The meat scraps are pressed to separate out the fat; the fat is cooked and turned into tallow and inedible fat, while the meat scraps are converted into a product called meat and bonemeal, a 50% protein material that is sold to feed mills as an animal feed concentrate.

The meat pressing is very odorous. The odors are collected by ductwork and forced into a counterflow scrubbing unit that stands approximately two stories high. The unit has a computer-designed air-to-water contact surface that is constantly rewet by a mist of sodium hypochloride ($NaHC10_3$) at the top of the tower. The concentration of hypochloride is such that 150 ppm of free chlorine is available.

The rendering plant uses some 250,000–300,000 gpd of water, primarily in the heat exchange system. About 4000 g/min of water are needed to cool the condensed steam from the cooking operation. Though the majority of this water is cooled in two large çooling towers and reused, some 10% is bled from the system for the steam condensing unit and the air scrubbing system. This water is sewered.

In the early 1970s, the grease from the first cooking operation was skimmed from the cooking solution and combined with the fats from the kill floor, and cooked separately to form a Number 2 tallow. These solids from the tallow processing were hauled out to agricultural lands.

The residual cooking solution was mixed with ferric sulfate, which lowered the pH to 5.5, causing the coagulation of the blood serum. The waste cooking liquor was then mixed with lime, to raise the pH to 6.5 in order to meet municipal treatment facility requirements. The solution next was fed to a flotation device that was equipped with a large number of DC electrodes and compressed air nozzles. The combination of air-saturated water and high concentration of DC current produced and efficient flotation process that removed most of the solids, forming a sludge containing about 6% solids (94% water). This sludge was difficult to dewater, but rich in protein and hence with a high nutrition content. A portion was heat dried to convert it to an animal feed; the remainder, five or six tankers daily, was hauled to agricultural lands for disposal during 9 months of the year.

Because of a variety of difficulties, mainly an inability to meet BOD discharge requirements and high sludge disposal costs ($750–$850/day in 1990), the process was modified. The cooking solids are now sent directly to the rendering system. The flotation system was removed and replaced by an anaerobic contact process in the late 1980s.

Most of the plant wastewaters are now treated by a mesophilic anaerobic contact process. Included are the following wastewaters:

1. Cutting floor washwater
2. Paunch washwater
3. Paunch press filtrate
4. Tripe processing liquids
5. Foot processing liquids
6. Hide salt washwater
7. Rendering skimmer wastes
8. Rendering floor washwater
9. Base flow (pump flushing and washwater)

Table 13-1 summarizes the overall influent wastewater characteristics.

After preliminary treatment in the existing skimmer, the wastewater streams are collected in an equalization tank, which provides flow and organic load equalization to ensure a uniform feed to the anaerobic reactor. From the equalization tank, the wastewater is pumped through a spiral heat exchanger to recover waste heat

TABLE 13-1.   Influent Wastewater Characteristics

Parameter	Concentration (mg/L)	Load @ 800,000 gpd (lb/day)
COD	7500	50,300
BOD5	3370	22,600
TSS*	2700	18,100

*TSS = total suspended solids

from the treated effluent and increase the temperature of the influent to approximately 30°C. Following the heat exchanger, direct steam injection is used to increase the temperature to about 34 to 37°C.

The anaerobic reactor is an above-ground, mechanically mixed, covered and insulated welded steel tank, in which microbes are used to convert the organic matter to methane and carbon dioxide. The system was designed for biogas production at an average rate of 0.24 standard m^3 of methane per kg COD added and an average biogas purity of 82% methane, and biomass production at an average rate of 9 kg total suspended solids produced per 100 kg COD added. This results in the following average removal efficiencies:

COD   84%
BOD5   93%
TSS   75%

The contents of the anaerobic reactor overflow to a degassification and flocculation tank. The biogas is stripped to ensure minimal sludge gassification and float formation in the subsequent settling stage. Baffles force the liquor to near the bottom of the tank, where a combination of shear forces and hydraulic head releases the gas from the bacterial floc.

The overflow from the degassification and flocculation tank is treated by an inclined plate settler, which provides liquid/solid separation. The underflow biomass solids are returned, as necessary, to maintain an adequate solids inventory. The excess biomass solids are conditioned with a food grade polymer and then dewatered using a belt press. The dewatered biomass is stored and subsequently rendered. The belt press filtrate is combined with the influent for recycling to the anaerobic reactor.

The biogas (methane) is burned as fuel in the boiler, as a replacement for natural gas. Variations in gas production and boiler demand are accounted for by a surge vessel and an intermediate storage tank. Any excess gas is flared. All available biogas is burned until the boiler demand exceeds the fuel supply, when natural gas is blended without interruption to supply the shortfall. Oxygen trim on both fuels compensates for variations in the heating value of the biogas.

The air emissions from the anaerobic system are also ducted to the air scrubbing system, for odor removal. Excess wastewater, which cannot be recycled, is sent to the municipal treatment facility.

SAUSAGE MANUFACTURE

Some of the meat from the packing houses is sold directly to retailers: some is processed into sausage or perhaps canned or cured meats. Sausage operations can be a part of the packing house operations. Frequently, however, the processing is done by independent companies (Figure 13-6).

The Process

Sausage is made of pork, ham, beef, bacon, veal, chicken, game, and sometimes fish. The meat is flavored with a variety of herbs and spices, including salt, black and red pepper, sage, garlic, onions, sugar, and ginger. The meat typically arrives at the sausage company boneless. Before use, it may be necessary to trim the meat to remove some fat and sinews. Depending upon the type of sausage desired, the meat is ground either finely or coarsely. Then it undergoes "emulsification" (Figure 13-7). The meat is chopped very finely, mixed with the various herbs and spices,

FIGURE 13-6.   The interior of a sausage company kitchen. (*Photo courtesy Progressive Sausage Company.*)

FIGURE 13-7.   The emulsification procedure. (*Photo courtesy Progressive Sausage Company.*)

and converted to a fairly thin paste. The purpose is to encapsulate the fat particles by proteins. Any preservatives, such as nitrates or nitrites, are also added at this stage. It may be necessary to intentionally cool the meat during this process to prevent the growth of harmful bacteria. This cooling process is done using very fine chips of ice.

After emulsification, the meat is extruded into the casings (Figure 13-8). These casings are frequently manufactured from natural intestines, preferably of hogs, cattle, or sheep. Artificial casings, of cellulose materials, can also be used.

There are three general types of sausages: fresh, cooked, or smoked. If fresh sausages are desired, no further processing is necessary. Cooked and smoked sausages require several additional steps. Cooked and smoked sausages are treated fairly similarly except, of course, for the smoking itself. The sausages that are to be smoked are placed on racks, and then conveyed to the smokehouse (Figure 13-9). In the smokehouse, they are heated to a minimum of 137°F (to prevent botulism): typically, temperatures around 150–160°F are used. The temperature and humidity are closely monitored within the smokehouse. Simultaneously, the sausages are

FIGURE 13-8.   Extrusion of the sausage meat into the casings. (*Photo courtesy Progressive Sausage Company.*)

subjected to a stream of smoke, usually produced by the burning of hardwood sawdust. The creosols are what give the meat the smoky flavor. Some companies instead spray the sausages with a liquid smoke extract.

After the smokehouse and cooking operations, the sausages are subjected to a hot water shower and then cooled. They are then ready for shipment to the retailer.

Environmental Considerations

The major wastes from sausage manufacture are fats, some protein, and a little blood. The initial trimming and grinding operations create some of these wastes, and the hot water shower utilized after the cooking and/or smoking operation generates a large quantity of grease. These wastes all are washed into the effluent water from the plant.

FIGURE 13-9.   The smokehouse operations. (*Photo courtesy Progressive Sausage Company.*)

The sanitation procedures themselves produce substances that must be disposed of. The smokehouse walls get coated with the creosols used in the smoking. They must be scrubbed daily, or, at the least, weekly (for the smaller companies that smoke only a small quantity of sausages). Caustic soaps are required to cut these creosols. A foaming agent may also be used. This generates a very strong waste.

The entire sausage plant must be washed down with large quantities of water, generally twice a day, followed by treatment with a sanitizing agent. According to Marvin Karrels, former owner of Progressive Sausage Company, a small independent firm: "The environmental problems [of the sausage companies] are considerably fewer than those experienced by packing houses. The wastes are primarily grease and trimmings. Sausage companies do not have to worry about the large quantities of blood or paunch wastes. The BOD load is thus considerably less."

Furthermore, according to Robert Risch, former president of Progressive Sausage: "Though some of the larger sausage companies, particularly those associated with packing plants, do treat their own wastes, the majority of the smaller independent companies, such as ours, utilize the municipal treatment facilities. The wastes tend to be very compatible with the normal secondary treatment procedures, and create few if any problems."

In addition to the above water pollution problems, the smoking operation will also create some air pollution. The excess smoke from the burning of the hardwood sawdust, that smoke is not deposited on the sausages or on the smokehouse interior walls, will be exhausted to the atmosphere and should, for larger operations, be collected by some scrubbing device.

CHEESE PRODUCTION[17]

The dairy industry encompasses many processes, including fluid milk production, creamery butter production, cheese manufacture, ice cream manufacture, and condensed and evaporated milk manufacture. The majority of dairy plants are small and scattered primarily over the milk-producing areas of the country. They range from single-product to multi-product facilities. This section will discuss only one type of dairy process, cheese manufacture.

The cheese-making process starts with the receipt and storage of the raw milk (Figure 13-10). Before pasteurization, the cream is separated out if the product is to be a low-fat cheese such as mozzarella. The cream can be used for the production of some byproduct.

The raw milk is then pasteurized by heating to 142–145°F for about 30 minutes, after which it is cooled and pumped into cheese vats. The milk is next inoculated with a culture, which causes the formation of curd. The remaining liquid is "whey," which can be used for byproduct manufacture, or can be wasted.

FIGURE 13-10.   An early method of milk transport (still used in some locations, as this 1977 scene in Ireland illustrates).

The curd, after rinsing, is salted, and may be cut or milled. Fresh, salted cheese "curds" are a favorite with many people for snacks, cocktail parties, or as a mild cheese for cooking. The curds are pressed and placed into can-shaped molds called "hoops" (Figure 13-11). These cheese hoops are stored in controlled-environment rooms for aging, the storage time being dependent upon the type of cheese being produced. It may be as long as several months.

After aging, the cheese can be directly packaged, or it can be converted to processed cheese. Processed cheese is formed by grinding the hoop cheese and then mixing it with stabilizers and various flavorings. The blended ingredients must be again pasteurized before packaging. After packaging, the cheese can be stored in a cold storage area until needed.

Environmental Problems

Almost 90% of the dairy plants in the United States are connected to municipal sewer systems.[18] Many are small and located near municipalities; the wastes

FIGURE 13-11.   Testing the curd production.

FIGURE 13-12.   Until recently, many cheese factories emitted their effluent directly into nearby streams. This was then a common sight. (*Photo courtesy Wisconsin DNR.*)

generated also tend to be compatible with municipal waste water treatment procedures. The whey, the curd wash water (containing whey), any product spillage, and cleaning water and soaps are the only significant wastes for all of the cheese processes. The whey is by far the greatest waste problem, not only because of its volume (several billion pounds are created each year), but also because of its high protein content and high acidity.

Whey is a dilute solution of lactose, protein, salt, and fats. Approximately one gallon of whey is produced for every pound of cheese. More and more whey is being converted to food and food supplements, rather than being used as animal food or wasted. Substantial growth in the cheese industry, together with the increasing acceptance of "synthetic" foods (such as imitation ice cream and cake mixes) and stricter pollution regulations, have led to larger and larger markets for whey products.[19]

Whey Recovery Techniques
Whey is a greenish-yellow fluid. It can be classified as "acid" or "sweet." The sweet whey (pH 5–7) is produced by whole milk, used in the production of natural or processed cheeses such as cheddar. Acid whey (pH 4–5) is produced by skim milk, often used to produce cottage cheese.

The BOD from a cheese plant producing 100,000 lb of whey daily is equivalent to that of a city of 22,000. Many cheese plants are smaller, but (particularly the

FIGURE 13-13.   This 1971 photo of a cheese factory outfall illustrates the contamination that was until recently emitted to our streams. (*Photo courtesy Wisconsin DNR.*)

small plants) are often located in rural areas with only minimal treatment facilities that cannot handle the whey properly.

About one-third of the 22 million lb of whey produced annually is used in human food or as animal feed supplements. Thus, about 14-3/4 billion lb annually has to be disposed of.[20]

In many instances, the cheese factory wastes are disposed of by irrigation techniques: spray fields (Figure 13-14), spreading (Figure 13-15), and ridge and furrow application (Figures 13-16 and 13-17). In order to utilize these methods effectively, the plant must own or lease sufficient land so that there will be no runoff to nearby streams. Leaching is a common problem (Figure 13-18), yet the land must be located sufficiently near the plant so that the wastes can be disposed of economically. Pipeline or truck are the common waste transfer methods. Odor problems can also develop, so it is preferred that the disposal areas be somewhat isolated. The nutrient value of the wastes makes it quite suitable for farm disposal in some locations.

Whey desalination is necessary before it can be used in other products. This desalination, or "de-ashing," as it is called, can typically be accomplished in several steps. The first is concentration of the solids from about 6% solids to 30% solids. Reverse osmosis can be used to preconcentrate the whey to about 15% solids, after which conventional dairy evaporation equipment can concentrate the whey to 30% solids or greater.

The pH of the whey is usually fairly high (except in the manufacture of cottage cheese). It is typical for it to be adjusted at this stage, and the insoluble proteins are removed by clarification.

FIGURE 13-14. Spray fields for disposal of cheese factory liquid wastes. (*Photo courtesy Wisconsin DNR.*)

FIGURE 13-15. Land spreading of liquid wastes. (*Photo courtesy Wisconsin DNR.*)

320

FIGURE 13-16. Cheese factory ridges and furrows system prior to irrigation. (*Photo courtesy Wisconsin DNR.*)

FIGURE 13-17. Ridge and furrow system during irrigation. (*Photo courtesy Wisconsin DNR.*)

FIGURE 13-18.   Leaching of a ridge and furrow system into a nearby stream. (*Photo courtesy Wisconsin DNR.*)

Electrodialysis can then be used to remove 25–90% of the salt from the whey, either by a batch or continuous process. Either before or after the electrodialysis, some of the lactose can be removed from the whey by crystallization; a warm solution of the whey is cooled. Up to 50% of the lactose can be removed in this manner.

The final step is the spray-drying of the liquid to produce a stable, free-flowing powder. As such, the whey has a long shelf life (as a fluid, the whey should be used within 24 hours).

A typical 90% desalinated whey is 80% lactose, 14% protein, 4.5% moisture, and less than 1% each of butter fat and salt. Its vitamin content is similar to that of nonfat dry milk. Often it is used in infant formulas and as a substitute for nonfat dry milk in ice cream, ice milk, sherbets, soft-serve products, and candy. The partially delactosed whey is used in protein drinks, dietary products, cereals, and baked goods.[21] In spite of these potential treatment methods and byproducts, the whey still is a major problem for many cheese factories. Although the sweet whey powder is a salable product, its production is not profitable. Furthermore, these treatment methods also have their limits and drawbacks. And when considering such huge quantities, the water discharged is still quite contaminated.

There have been several studies funded in recent years to develop better whey treatment methods. One of the new developments involves the use of membranes, but in a manner different from the traditional procedures. This procedure consists of two stages—ultra-filtration and reverse osmosis.

The ultra-filtration, through about a 20-Å pore size membrane at about 60 psi and 120°F, removes the proteins, and can concentrate the whey by a factor of about 20. The deproteinized solution then goes to a 90°F high-pressure reverse osmosis system for lactose concentration, and a lowering of the BOD by up to 97%. The reverse osmosis membrane typically has a 3Å pore size and operates at 800–1000 psi.[22]

This process shows promise both environmentally and economically, though new markets for whey powder and other byproducts still need to be developed.

The Kraft Foods Division Facility at Stockton, Illinois[23]

The Kraft Foods Division facility located at Stockton is a good example of the type of treatment and disposal methods frequently utilized by cheese factories. The Stockton facility really consists of two separate plants: one for the manufacture of Swiss cheese and the other for the processing of whey. The whey processing facility not only treats the whey generated in the Swiss cheese production, but also the whey formed at other nearby Kraft cheese plants.

The Stockton plant has attempted over the years to reduce the BOD content of the water discharged. For example, in the period 1968–1972, the total BOD load decreased from 1950 lb/day to 900 lb/day, though the total water discharged increased from 86,000 gallons/day to 110,000 gallons/day. This BOD reduction was accomplished by some in-plant process modifications and improved housekeeping techniques.

During the summer, the total discharge is sprayed over a land area of about 50 acres. Thirty-two sprayheads operate in rotation. Each sprayhead pumps 90 gallons of waste/minute and individually irrigates a 220-ft diameter circle. Automatic controls prevent overdosing, and maximize the percolation, evaporation, and transpiration of the liquid (Figure 13-19).

During the winter, the wastes are treated by an activated sludge system. After treatment, the discharge water is disposed of by an on-site ridge and furrow system. The sludge is taken from the final clarifier to a nearby sludge lagoon, from where it is removed once a year and spread on adjoining grassland (Figures 13-20 and 13-21).

In order for the Stockton plant to discharge any of its wastes into a nearby stream, it would have to meet very strict Illinois effluent concentration requirements. Rather than attempt to meet these requirements, it was decided to instead depend on evaporation and percolation for on-site disposal. This meant that the disposal sites had to be well protected against floods. Consequently, two large flood control

FIGURE 13-19.   Sprayheads used by Kraft for land disposal of whey—containing waste water. (*Photo courtesy Wisconsin DNR.*)

ponds were located at the lowest point on the property to catch any runoff from either summer or winter operations.

Chemical Recovery[24]

As is apparent, many food processing firms are burdened with large volumes of food waste byproducts, including potato peelings, cornstarch, and similar materials. For example, an estimated 10 billion lb/year of potato waste is produced solely by firms making ready-cut french fries and other processed and unprocessed potatoes. Though many of these wastes can be sold to farmers for cattle feed, this practice is only marginally profitable.

Some food processing companies have sold their wastes for conversion to ethanol (mainly for mixing with gasoline to produce gasohol) or methane. Though the number of gas stations selling gasoline-alcohol mixtures has risen dramatically since the mid-1970s, this has not developed into a major market.

To maximize the profits that potentially could be realized from the high carbohydrate concentrations in these wastes, a number of research facilities (including Argonne National Laboratory and Battelle Columbus Labs) are developing techniques to convert the starch in potato peelings and cornstarch waste to lactic acid, with an ultimate goal of inexpensively producing polylactic acid. Polylactic

FIGURE 13-20.   Whey pits. (*Photo courtesy Wisconsin DNR.*)

acid is a naturally degradable plastic that will degrade in landfills. Because of its high cost when made from non-waste raw materials and using existing technology, at the present time it is used only for speciality products, mainly in the human body for sutures, wound-clips, or drug delivery systems.

To produce the polymers from waste starches, recently developed biotechnology is used. Typically, the starch is homogenized in a blender and then subjected to two enzymes. A high-temperature alpha amylase solubilizes the starch; glucoamylase then breaks apart the polymeric chain structure, converting > 90% of the starch to glucose. The glucose can be fermented into lactic acid by a bacterium, lactobacillus.

Though lactic acid is a salable product, any significant use of waste food starches would soon flood the market. Some 20 or 30 million lbs/year of potato starch, for example, is available. However, the lactic acid can be polymerized to polylactic acid (PLA), which potentially could have a much greater market.

The raw materials are very inexpensive (potato peelings cost about 1 cents/1b and cornstarch about 5–9 cents/lb), but existing polymerization processes are quite costly. Molecular weights around 50,000 g/mol are needed to develop useful physical properties. Self-condensation, though possible for lactic acid, is normally

FIGURE 13-21.   Whey disposal area. (*Photo courtesy Wisconsin DNR.*)

limited to molecular weights near 10,000 g/mol due to the onset of decomposition and chain termination. Research, therefore, has focused on the development of indirect polymerization techniques that are less expensive than the current processes.

The likely application of these waste-derived plastics will be primarily for agricultural purposes, such as for time-released coatings for fertilizers and pesticides and tillable, agricultural mulch films to maintain heat and moisture and to reduce weed growth.

References
1. "Processes Cut Canning Pollution." *Environmental Science and Technology* **7**, *No. 10*, October 1973, p. 900.
2. *Ibid.* p. 900.
3. *Pollution Control Technology.* New York: Research and Education Assôciation, 1973, p. 452.
4. Jones, R. H., *Waste Disposal Control in the Fruit and Vegetable Industry.* Park Ridge, New Jersey: Noyes Data Corporation, 1973, p. 27.
5. *"Processes Cut Canning Pollution. "* p. 900.
6. Hoover, Sam R., "Prevention of Food Processing Wastes." *Science* **183**, March 1974, p. 824.
7. "Processes Cut Canning Pollution." p. 900.
8. *Ibid.* p. 901.

9. *Ibid.* p. 901.
10. Krofchak, David and Stone, Neil J., *Science and Engineering for Pollution Free Systems.* Ann Arbor: Ann Arbor Science Publishers, 1975, p. 276.
11. Jones. p. 156.
12. Jones, R. H., *Pollution Control in Meat, Poultry and Seafood Processing.* Park Ridge, New Jersey: Noyes Data Corporation, 1974.
13. Macon, J. A. and Cote, D. N., *Study of Meat Packing Wastes in North Carolina.* North Carolina State College Ind. Extension Service, 1961.
14. Bethea, Robert *et al.*, "Odor Controls for Rendering Plants." *Environmental Science and Technology* **7**, *No. 6*, June 1973, p. 504.
15. *Ibid.* p. 509.
16. *Ibid.* p. 509.
17. Jones, R. H., *Pollution Control in Dairy Industry.* Park Ridge, New Jersey: Noyes Data Corporation, 1974.
18. Watson, Kenneth, *The Treatment of Dairy Wastes.* Prepared for Environmental Protection Agency Technology Transfer Program. Philadelphia, Pennsylvania, August 21–22, 1973, p. 2.
19. Leitz, B. Frank, "Electrodialysis for Industrial Water Cleanup." *Environmental Science and Technology* **10**, *No. 2*, February 1976, p. 136.
20. "Membrane Processing Upgrades Food Wastes." *Environmental Science and Technology* **5**, *No. 5*, May 1971, p. 396.
21. Leitz. p. 138.
22. "Membrane Processing Upgrades Food Wastes." p. 137.
23. Watson. p. 6.
24. Keeler, Robert, "Don't Let Food Go to Waste—Make Plastic Out of It." *R & D Magazine*, February 1991, pp. 52–57.

14

The Brewing Industry

THE PROCESS

The brewing industry involves the making of fermented alcoholic beverages, such as beer and ale, from cereal grains. There are two major steps involved in the process: malting and brewing.

Malting

The purpose of the malting is to prepare the grain for brewing. Wheat, corn grits, or barley are soaked in water for 48–76 hours. The water is aerated, drained, and replaced at least once a day.

When the grain is soft, it is piled in heaps on the "couching floor," or placed in germination bins, where it is kept at constant temperature and humidity for about a week. The grain germinates (i.e., sprouts small root shoots) and, simultaneously, the crude starch in the grain is converted to soluble sugars and starches by enzymatic action. As the germination proceeds, the grain is turned intermittently to assure even sprouting.

When the sprouts are about two-thirds the length of the grain, the germination is halted. The grain is transferred to drying kilns, where it is gently toasted at 155–220°F until it is dark and crisp. Lower temperatures produce a light beer; higher temperatures, a dark beer. This dry-roasted grain is now called malt.

Brewing

The dry malt is then crushed with iron rollers and mixed with water to form mash. The mash is thinned with hot water and heated to 145°F while being stirred constantly. The exact temperature is important at this stage, for the malt is undergoing chemical change; the remaining starches are being converted to malt

FIGURE 14-1.   Aerial view of the Miller Brewing Company plant in Fulton, New York.

sugar. After a while, the temperature is slowly increased to 160°F. Then the cooked mash is filtered, and the liquid, called the wort, is drained from the grain, or the grist.

The wort, a clear amber malt extract, is boiled 1-6 hours with hops, which add flavor and aroma. Hops are the dried flowers from the hop vine, and themselves possess a bitter flavor. For every 100 gallons of wort, 1-12 lb of hops are added. As well as adding flavor, they also help keep the wort from spoiling.

The spent hops are removed, and the wort allowed to cool. As it cools, the undesirable proteins coagulate and settle out.

The wort is then placed into fermenting vats (Figure 14-2). About 5 lb of yeast are added per 100 gallons of wort. Two fermentations occur. Within the primary fermentation tanks, the malt sugars are converted into alcohol and CO_2. Beer undergoes "bottom fermentation" at 43-46°F. The yeast sinks. Ale, porter, and stout undergo "top fermentation." The yeast floats on the wort, which is maintained at 60-68°F. After several days, the fermented liquid is poured into settling vats. Secondary fermentation, all the reactions required to give the fine, distinctive quality to the beverage, then takes place. The yeast is skimmed off or the beer drawn off, and the beverage is stored in casks and barrels to age and become clarified. After a short period, the finished beer, ale, or stout is ready to be bottled.

FIGURE 14-2.   The brewing kettles at Pabst Brewing Company. These solid copper kettles have been in use since the late 1800s. (*Photo courtesy Pabst Brewing Company.*)

ENVIRONMENTAL PROBLEMS

General Pollution Problems

Some of the settled yeast from the previous fermentation is used to initiate fermenting in the next batch. There is always excess yeast, though; not only does too much develop, but only part of the yeast is satisfactory for further use. The settling process carries down some materials not desirable in these beverages. This limits the useful quantity. There is always a fairly large amount of yeast that must be disposed of.

The residual yeast is high in Vitamin B complex, proteins, and minerals. If the excess yeast is discharged to the sewer, it is objectionable because of the high BOD. It can, instead, profitably be dried and used as cattle and poultry feed. The drying deactivates the enzymes, stabilizing the yeast, and also improves its digestibility.

It is estimated, however, that 1,000,000 lb of yeast solids/year are necessary to economically justify the installation of drying equipment.[1]

The brewing industry consumes much water—about 10 gallons of process water/gallon of product. The BOD levels are quite high, as are the total solids. Typically, about half the BOD and over 90% of the suspended solids are generated in the brewing operation.[2] There are also solid wastes—spent grains, hops, and sludges—that are formed in this and the malting steps, and that must be disposed of. Packaging operations also generate a large amount of BOD.

A typical treatment system employs an activated sludge process. After mechanical bar screening and grit removal, the wastes are sent to an aerated equalization basin. At that point, various chemicals (such as ammonia and phosphoric acid) are added for pH control and nutrient addition. The liquid wastes are next sent to primary clarifiers, and then to the activated sludge basins and the secondary clarifiers. The waste-activated sludge is thickened, and then combined with the primary sludge, dewatered, and disposed of by landfill procedures.

Sludge Bulking[3]

The growth of the bacterium *Sphaerotilus natans*, plus that of a few other microorganisms, leads to a phenomenon generally experienced in the brewing industry and several other industries: sludge bulking. A bulking sludge is one that settles slowly (leaving a clear supernatant), has a high volume, and compacts poorly. It generally leads to a major discharge of solids into the effluent.

Though a number of other microorganisms can contribute to the problem, the majority of the bulking is generally attributable to *Sphaerotilus*, whose filaments are composed of chains of rod-like cells with rounded ends that are encapsulated in a tight polysaccharide sheath. The *Sphaerotilus* thrives at temperatures of 15–40°C, and at pH values between 6.5 and 8.1.

Within the brewery there are several probable causes of sludge bulking. They include the following[4]:

1. *Low levels of dissolved oxygen.* Dissolved oxygen levels in the mixed liquor of less than 1.0 mg/L promote the growth of the filamentous *Sphaerotilus* organism. Specifically, the growth of *Sphaerotilus* is retarded less by low dissolved oxygen levels than is the growth of other types of organisms, those necessary within the activated sludge, thus allowing it to predominate.
2. *Loading rate.* The filamentous organisms will most likely grow if the BOD level is within a certain range.
3. *Nutrients.* Insufficient quantities of nitrogen, phosphorus, and iron promote bulking. Typically, a BOD-to-nitrogen-to-phosphorus ratio of about 100-to-5-to-1 is desirable.
4. *Waste composition.* Wastes high in carbohydrates promote the development of

filamentous organisms, possibly due to the rapid degradation and the low dissolved oxygen levels that frequently result.

5. *Shock loading*. Any sudden changes in the flow rate or similar factors may contribute to increased bulking. This may also be due to the occurrence of low dissolved oxygen.

6. *pH*. A low pH in the aeration basin (below 6.0) promotes the growth of many types of filamentous fibers.

What can be done to control the bulking in activated sludge systems? Many techniques have been attempted with varying success. Most approaches involve: 1) killing the filamentous organisms by chemical addition, 2) increasing the sludge's settleability by using chemical coagulants, and/or 3) recirculating the sludge more, to prevent buildup.

The use of chemical additives is based on the principle that the long, filamentous organisms, with their large surface-to-volume ratios, are more sensitive to various disinfectants than are the normal activated sludge microorganisms. Typical chemicals used include chlorine and hydrogen peroxide. It is believed these can oxidize some of the filaments, destroying the lattice-like framework; hydrogen peroxide can also provide additional oxygen.

Chemical coagulants do not affect the number of filamentous organisms, but simply try to improve the settling in the secondary clarifiers. Frequently, cationic organic polymers or iron or aluminum salts are used.

Filamentous organism buildup can also be reduced by process modifications. For example, the rate of return of the activated sludge can be increased to increase the percentage of high-density solids and thus decrease the overall volume of sludge. Likewise, more activated sludge can be wasted to decrease the overall volume. The success of these various methods depends upon the nature of the effluent.

Overall, maintenance of adequate dissolved oxygen is the most critical factor in bulking control. The presence of sufficient nitrogen and phosphorus are also essential. Simply maintaining stable operating conditions is very important, too. Any changes in recycling rates, sludge wasting rates, and so forth, must be made very slowly to minimize filamentous organism growth and to allow appropriate corrective action to be taken when necessary.

The Coors Brewery

An example of the type of pollution control efforts that are necessary in a brewery are typified by the Adolph Coors Company.[5] The Coors Brewery is located in Golden, Colorado, near Denver.

Until 1953, the effluent was discharged directly into Clear Creek, a tributary of the South Platte River. At that time, the Coors plant built a primary treatment facility to treat not only its own but also Golden's municipal wastes.

A series of screen and mechanical separators trapped the solid wastes, spent hops, grains, and paper labels. These solids were pumped to anaerobic digesters fitted with extra gas and steam recirculation systems to increase the mixing.

There were difficulties with the operation of the system due to the type of plant. The malting and brewing operations are batch processes: not only are the waste quantities high, but they also vary drastically. The pH also fluctuates significantly.

A two-pronged improvement program was necessary. First, a recovery system was installed to reduce the wastes, and also to recover valuable byproducts. For example, waste beer from the bottle-filling lines, instead of being shunted directly to the treatment plant, is collected and then cooked in evaporative condensers. This waste beer forms a syrup that can be used as a binder for the production of pellets of spent grains and hops, to be used as cattle feed.

Secondly, an activated sludge process was installed. Due to the high-carbohydrate effluent, the traditional process had to be modified. The carbohydrates promoted the rapid growth of the filamentous *Spherotilus natans*, which would otherwise clog sedimentation basins. The system was modified to include the injection of small amounts of highly digested sludge directly into the activated sludge return. This minimizes the plant formation.

The wastes from the Coors plant also do not contain adequate nitrogen and phosphate for the process to operate successfully. The Golden city wastes have supplied adequate phosphate, but it is necessary to add anhydrous ammonia to supplement the nitrogen.

The clarified effluent is chlorinated and then discharged to Clear Creek. The waste-activated sludge is thickened by flotation, mixed with the primary settled sludge, and then further thickened by vacuum filtration. The dewatered sludge has been used as a soil conditioner on company land.

References

1. *Pollution Control Technology.* New York: Research and Education Association, 1973, p. 459.
2. Becker, Kenneth, Schwartz, Henry G., Jr., and Popowchak, Theodore, "Sludge Bulking in the Brewing Industry." Presented at the Annual Conference, Water Pollution Control Federation. Anaheim, California, October 2, 1978.
3. *Ibid.*
4. *Ibid.*
5. "Reuse, Recovery Lower Pollution from Brewery." *Environmental Science and Technology* 6, *No. 6*, June, 1972, p. 504.

15

The Tanning Industry

THE PROCESS

The tanning process is undertaken to convert the normally putrescible proteins in an animal skin to some stable form, to preserve that skin for leather or fur pieces. Leather differs from the raw hide also in terms of its greater dimensional stability as to changing temperature and humidity, as well as its greater toughness, flexibility, and lower water permeability. The tanning process does, however, retain the original morphological structure of the skin.

All types of skins can be tanned, but cattle hides predominate, followed by those of calves, sheep, deer, and goats. Smaller numbers of skins of other animals, such as ostriches and lizards, are also converted to leather.

When the skin or hide arrives at the tanning plant, it consists of two layers: 1) the flesh, which is really not part of the skin, and 2) the derma.

Varying amounts of flesh are attached, depending upon both the type of skinning process employed at the meat packing plant and the skill of the flayer who did the skinning. This flesh is removed in the early stages. It consists primarily of muscle tissue, fat cells, blood vessels, and adipose tissue.

The derma is the part of the skin that actually becomes leather. It, too, can be subdivided into two layers: 1) the epidermis, and 2) the corium.

The epidermis is the top layer. It contains the hair, the hair follicles, and the oil glands that surround the hair roots. On the surface of the epidermis is a thin layer of a non-leather-making protein that must be removed. Running through the epidermis is a network of collagen. This collagen is the actual leather-making protein. In the early tanning stages, all the undesirable materials are removed—substances such as keratin, elastin, mucoids, albumens, fat, and other proteins and alkaloids.

The corium is the major part of the derma. It consists primarily of the protein collagen, in the form of interwoven fibers and fiber bundles. The objective of the tanning is to cleanse the skin of as much unwanted "debris" as possible, and then to process the collagen fibers to render them strong, flexible, and nonputrescible.

The chemistry of the skin is thus primarily the chemistry of proteins. Regardless of the type of tanning process, in the early stages many of the undesirable proteins are hydrolyzed or otherwise broken down and removed. Many of these substances become colloidal or dissolved, and lead to water pollution. Many of them are not very susceptible to decomposition by microorganisms, and hence are difficult to treat by conventional secondary processes. Special enzymes are frequently required.

Preparation of the Hides and Skins

The skins and hides are received from the meat packer or from other sources such as hunters, in a "cured" form, (i.e., treated to prevent rotting before they reach the tannery). Usually, they are in bundles, salted or brined for preservation. They may have been first "prefleshed," or processed through a machine to remove the majority of the flesh. They had probably then been immersed in a circulating saturated sodium chloride brine, which also cleanses the hide of its manure, dirt, and blood. The sodium chloride absorbs water, and thus dehydrates the proteins, making them less susceptible to spoilage.

At the tanner, the hides are split by cutting down the backbone to form two "sides." These sides are immediately soaked to remove the curing agents and to restore the moisture to render them flexible and receptive to chemical treatment. The hides are placed into large vats equipped with paddle wheels. A wetting agent may be added to the water. This initial soaking may last 2–48 hours, depending upon the hide. Small quantities of sodium sulfide and lime may be added halfway through this process.

After rinsing, the hides are fleshed. The skins are put through machines equipped with a rubber roller and a shaft to which spiral blades are attached. These blades cut the flesh and tissue from the inner surface of the skin, leaving a clean, uniform surface. The waste solids are collected in a box with holes for draining (Figure 15-1) and the liquids are sent to the sewer.

The hides are then transferred to the beamhouse for dehairing. The hides are soaked for 3–7 days in a saturated solution of lime that also contains a small amount of sodium sulfide or sodium sulfhydrate and possibly other chemicals. This solution and the hides are placed in a cylindrical drum and are rotated slowly to assure mixing. The hair is loosened and can easily be removed in a dehairing machine, which scrapes the hair off with a series of blunt blades. The solution in which the hides are soaked is taken to a holding tank to be refortified and reused. The hair is recovered, washed, dried, and baled for byproduct recovery. The washings that contain protein solids, grease, and some sulfides and lime enter the effluent stream.

The sides are then immersed in another lime suspension, with again a small

FIGURE 15-1.   Paddle vats used for soaking and unhairing operations. (*Photo courtesy New England Tanners Club.*)

amount of sodium sulfide, and agitated with a paddle. This lime swells the skin and opens the finely interwoven collagen fiber bundles. The non–leather-making proteins and further debris are removed. The flesh side of the hides becomes rough due to the swelling of tissues not removed in the first fleshing process, and the hides may be refleshed. The hide is then sent to a "scudding" machine, which scrapes the grain surface with a dull blade (Figure 15-2). This removes the last vestiges of hair roots, epidermal keratinous tissue, and other surface debris. This debris also enters the effluent stream.

Before tanning, the hides and skins must undergo "bating" to remove the lime (Figure 15-3). After liming, the hides are in a highly swollen, alkaline state. They must be made acidic, and the swelling removed.

FIGURE 15-2.   Cutting away unwanted fleshy matter on a fleshing machine. (*Photo courtesy New England Tanners Club.*)

FIGURE 15-3.   Drums used for bating, pickling, tanning, retanning, coloring, and fatliquoring operations. (*Photo courtesy New England Tanners Club.*)

The sides are placed in a drum and washed thoroughly. They are then treated with a proteolytic enzyme combined with a lime-neutralizing substance such as ammonium sulfate or ammonium chloride, at a pH of about 9.0. A small amount of detergent might be added. The enzyme reacts with a noncollagenous protein in the grain surface, changing its physical condition. The grain develops a silky feel, and the side becomes soft and pleasant. After 30 minutes to 4 hours, the bating is over; the hides are washed with cold water, to remove the lime as soluble calcium sulfate or calcium chloride, and piled flat to drain. The hide is then ready to be tanned.

Tanning

Though all tanning processes accomplish the same general result, and all involve soaking the skin in a tanning agent, the tanning agent associated with each process varies widely in chemical composition and rate of tanning. While all tanning agents combine chemically with the collagen, each produces a different type of leather and has its own advantages and disadvantages.

There are three types of tanning agents: vegetable, mineral, and organic. The vegetable tanning agents are substances such as aqueous extracts of barks and plant tissues, typically from oak, chestnut, wattle, and quebracho. The tannins from these substances are the active ingredients. Tannins are, chemically, phenol carboxylic acids and polymeric polyhydric phenols. They require up to several weeks to tan a hide, and are used to produce primarily leathers for shoe soles. Mineral tanning agents consist of substances such as chromium sulfate, alum, and zirconium. Chromium sulfate is the most widely used of the minerals. It acts rapidly, and is used primarily for thin leathers for shoe uppers and similar products. The alum and zirconium tans are used on furs. Organic tanning agents include easily oxidized fish oils, which are used to produce chamois, and a combination of formaldehyde and urea, used for special leathers. Naphthalene formaldehyde sulfonates frequently are used as supplementary tanning agents.

Some typical tanning procedures are described below.

Chrome Tanning

Chrome tanning is generally done on cow and steer hides to produce shoe uppers and similar products. It can also be used to produce fine-quality sheepskin coats.[1]

After bating, the hides are pickled for 4–8 hours—reduced to a pH of 2.0 in a sodium chloride solution with added sulfuric acid. The pickling neutralizes any remaining lime and dehydrates the hide fibers. The sodium chloride suppresses the swelling. The hides are then placed into drums containing a complex chromium salt, bichromate of soda, formed by reacting sodium bicarbonate ($NaHCO_3$) with sulfuric acid and an organic such as sucrose, or sugar ($C_{12}(H_2O)_{11}$). Also in the tanning solution are sodium chloride and sulfuric or hydrochloric acid. After soaking to allow the solution to penetrate the skin, the pH is raised to 3.5 by the addition of sodium bicarbonate or sodium thiosulfate ($Na_2S_2O_3$). At this higher pH, the bichromate of soda is reduced to either chromium sulfate or chromium chloride (depending on which acid is used), and the chromium can chemically combine with the collagen, forming leather. The leather is then washed in running water. The excess solution is drained and wasted.

The leather is wrung to extract the moisture, and then often split horizontally with a knife. Two pieces are formed—the grain or top layer and the split or bottom layer. The "blue," or "chrome," split is used for shoe linings, insoles, and slipper soles. The grain is sent through a shaving machine to smoothen it. The shavings are collected and then baled and sold as a byproduct. Some escape to the effluent.

Vegetable Tanning

Vegetable tanning is used to produce leather for shoe soles, handbags, straps, harnesses, upholstery, and belts. The tannin in the vegetation is the active ingredient. The principal sources of tannin are the leaves, nuts, woods, and barks from various trees and plants. It can generally be extracted by hot water.

The exact tanning procedure depends upon the desired end result of the leather. Sole leather is, for example, processed for up to 135 days. Bag, harness, and strap leathers, particularly when the hide is sheepskin or calfskin, can be processed in only a few hours.

Usually, tannins from several sources are mixed. Sole leathers may, for example, be processed by tannins from hemlock, chestnut, oak, and quebracho (an Argentinian hardwood tree), with small amounts of the extract catechu obtained from an Asiatic plant, and some myrobalan, a dried astringent fruit much like a prune, produced by tropical palm trees.

After the tannins are mixed, they are diluted as weak as 0.5–1%, and then placed in vats equipped with rockers to which the hides are attached. A gentle rocking motion is obtained. The hides are put into vats with successively stronger tanning solutions (perhaps 15–30 of these vats), with solution concentrations reaching 4%. This process takes 15–30 days. Then the hides are placed in stronger liquors for another 15–100 days. After this period of time, they are rinsed, placed in a tanning drum, and mixed with concentrated tannin to increase the durability of the leather.

Next, the hides are "tempered" by soaking in warm, strong tanning liquor for 3–5 days. After draining overnight and being wrung to remove all liquid, they are ready for further treatment, including splitting and shaving.

The tanning process can be sped up through the use of hexa-meta-sodium phosphate $(NaPO_3)_6$. If the hides or skins are treated with this chemical after bating, heavy sole leather can be tanned in only 6–7 days.

Further Processing

Regardless of the tanning procedure chosen, the leathers all must undergo some further treatment.

Dyeing

After splitting and shaving (Figure 15-4) the leather is dyed. First the leather is rinsed and then placed in closed drums with a number of dyes. More than one type is always necessary to attain the desired color, durability, depth, and fastness of color. These dyes include tannin, acidic and basic aniline dyes (based on aniline, $\langle \cdot \rangle$–NH_2), and natural wood dyes. The dyes may be "fixed" with formic acid (HCO_2H). The dyeing time varies between 1 and 3 hours. After the dyeing, the leather is fat liquored.

FIGURE 15-4.   Splitting side leather to uniform thickness. (*Photo courtesy New England Tanners Club.*)

Fat Liquoring

All leathers are treated with oil or grease so that the fibers are flexible and strong. Various oils are put into water emulsions and then worked into the leather. The oils used are typically cod, sperm, or other animal or fish oils. Usually, they are sulfated to make them self-emulsifiable, but additional emulsifiers and stabilizers could also be added.

Drying and Setting

There are several methods used to finally dry the leather. It can be pasted or affixed to smooth porcelainized steel plates sprayed with water-dispersible paste, and then passed into a heated dryer (Figure 15-5). After drying, the hides are stripped off and the old paste is sent to the sewer.

The leather could also be pressed in machines to remove excess water, iron out the wrinkles, and smooth the grain, and then hung to dry. After 3–5 days hanging in the atmosphere, the leather would be "sammied," or dipped in warm water for 1 or 2 minutes, and then placed in damp sawdust for 24–48 hours. After this, the leather would be tacked on boards or toggled on perforated metal frames to thoroughly dry (Figure 15-6).

Finishing

The finishing involves applying solutions or emulsions of pigments, waxes, oils, dyes, varnish, shellac, and similar materials to the grain surface of the leather. Patent leather is produced, for example, by applying three coats of a heavy, oily varnish. Between applications, it is necessary to roll, glaze, and smooth the leathers mechanically.

FIGURE 15-5.   Slicking wet skins on adhesive-coated plates for paste drying. (*Photo courtesy New England Tanners Club.*)

FIGURE 15-6.   Toggling wet skins on metal frame preparatory to drying. (*Photo courtesy New England Tanners Club.*)

POLLUTION CONTROL[2]

The tanning processes produce both solid and liquid wastes. The solid wastes are predominantly trimmings from both the initial and final steps, fleshings, hair, and shavings. The liquid wastes tend to be either basic or acidic. The basic effluent arises from the following operations: the initial soaking and washing, the fleshing, the de-hairing, the paste drying, and the finishing. The acidic effluent is produced in the bating, pickling, tanning, dyeing, and fat liquoring operations. Often, the acidic and basic wastes are collected separately. These wastes are summarized in Figure 15-7.

Continuous treatment of tanning wastes is usually impractical because of the wide fluctuations in the volume and quality of the wastes. Instead, the liquids are stored for 4–8 hours in an equalization or retention basin, from which they are released gradually. The acids are collected separately, then added as desired to retain the pH at about 9.5.

The major pollutants in the wastewater are typically calcium, sulfur, suspended solids, BOD, and alkalinity. Chrome tanning generates about 54,000 metric tons of trivalent chrome waste annually in the United States.[3] The primary sedimentation removes the bulk settleable solid, and further equalizes the waste quality. The sludge is typically greater than 8% solids, rather like a wet clay. Additional calcium can be removed by bubbling flue gas through the effluent. Burning oil, for example, produces flue gas with about 11% CO_2. The CO_2, in the presence of the tannery wastes, forms calcium carbonate, which precipitates.

The wastes can then be taken to an activated sludge process. Tannery wastes do generate a fair amount of foam, which has to be controlled by sprays or anti-foaming agents. As mentioned earlier, some wastes are hard to treat and require special

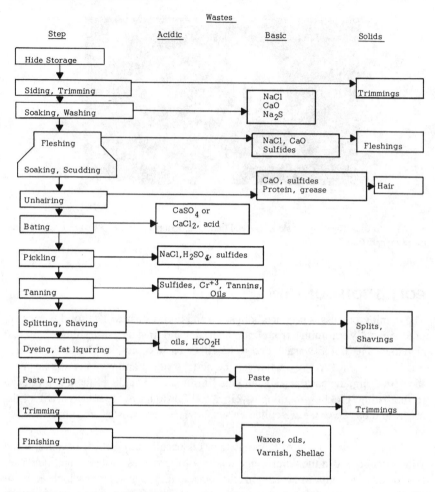

FIGURE 15-7.   Flowsheet for typical tanning process and wastes. (From *Pollution Control Technology*. New York: Research and Education Association.)

enzymes. The activated sludge may be light and difficult to settle and dewater. Mixing the sludge with the primary sludge by resettling it in the primary basin produces a mixture that is easy to handle.

Sludge dewatering can be done readily on drying beds, since there is no odor, but due to the large volume and high cost of the manual labor required for hauling, it is better to centrifuge the solids. The sludge can be increased to 20–30% solids, and then trucked to landfill. Landfills, however, are reluctant to accept the wastes if they contain trivalent chromium, due to possible conversion to hexavalent chromium chloride, as well as the subsequent seepage into groundwater.

Some of these wastes can be quite hazardous. There has been at least one reported instance where noxious tannery wastes were dumped into a Massachusetts sewage system, and the treatment plant had to shut down for several months because of the fumes, which overwhelmed the workers.[4] Proper care is always required within the plant and in disposing of the wastes.

Pfister & Vogel Tanning Co.[3]

Pfister & Vogel Tanning Co., located in Milwaukee, WI, is illustrative of the new pretreatment systems and requirements for, especially, chrome tanning wastes. The plant employs a caustic alkaline process using sulfides for hair and waste removal.

In 1971, Pfister & Vogel pioneered a chromium recovery process capable of recovering 64% of the trivalent chromium. This process was further modified and developed in subsequent years with the help of a 1974 EPA grant so that the effluent could meet the new EPA standards for chromium release, in effect February 1993.

After the tanning step, the chromium-containing water is wrung from the leather and processed for recovery of about 240,000 lb/year of trivalent chromium. The majority of the recovered chromium is stored in 26-ft tall vats, and then reused in the tanning process. The new $1 million recovery system, installed in January 1991, reduced the chromium discharge to the sewers from 113 ppm (without recovery) to 8–12 ppm. The new recovery system also generates an "excess" 20% trivalent chromium, of which 82% is converted to other usable chemicals, leading to a savings of $300,000. Processing chemicals cost about $100,000.

Pfister & Vogel is similarly attempting to convert the sulfides present in the waste stream to safe, reusable chemicals. The EPA has forbidden any sulfides in effluents after 1992, for they can produce toxic vapors and corrosive H_2S. They are, in addition, developing techniques to extract chemicals from scrap leather. Nitrogen-rich scrap leather can be stripped of some chemicals to produce animal feed and fertilizers. Collagen potentially can be extracted for use in suntan oil, and other proteins can be used to manufacture fire-fighting foam. These strategies can not only recover useful products, but can keep waste out of landfills.

References
1. Thompson, Norm, *Escape from the Ordinary.* Portland: Norm Thompson, Fall-Winter, 1978, p. 42.
2. *Pollution Control Technology.* New York: Research and Education Association, 1973, p. 434.
3. Wucker, Michele, "Chromium Cleaning." *Milwaukee Sentinel*, April 18, 1991, part 4, pp. 1–2.
4. "Hazardous Wastes: Hidden Danger." *Garbage Guide (RCRA Resource Guide No.2).* Environmental Action Foundation, 1978, p. 1.
5. Wucker. 4, pp. 1–2.

16

Chemical Industries

The chemicals industry can be divided into two general classes: inorganic chemicals and organic chemicals. Both of these are very large industries, and each utilizes its own types of processes and experiences, its own problems and difficulties.

In 1989, the first three "top-volume" chemicals produced in the United States, were inorganic (sulfuric acid, nitrogen, and oxygen, in order); in fact, inorganic chemicals secured eight of the top ten rankings, with the exceptions being ethylene, which was ranked fourth, and propylene, tenth. On the other hand, the production of organic chemicals overall has been gaining the larger volume. During the decade from 1979–1989, production of organic chemicals rose by 1.4%, whereas inorganic chemical production increased only 0.2%.[1]

THE ORGANIC CHEMICALS INDUSTRY

The organic chemicals industry is currently founded on one basic raw material: petroleum. There are only a few exceptions, those being foods, paper products, soaps, fermentation products, and perfumes.

At various times in recent U.S. history, often sparked by predicted petroleum shortages and prices, there has been renewed interest in the production of chemicals from biomass—resources such as coal, wood, cellulose, lignin, and starch.[2] A number of bio-based materials and chemicals are under development, including composites of conventional plastics with lignocellulosics (which can be pressed into rigid shapes to form walls, doors, and auto body parts), biodegradable plastics containing starch, and phenolic chemicals derived from wood waste and bark.

FIGURE 16-1.   Early chemical production facilities. (Apothecary Museum, Heidelberg, Germany.)

Large-scale adaptation of biomass raw materials are very dependent on the relative prices of petroleum and the specific form of biomass. Estimates are that (in 1990 dollars) at oil prices of $17/barrel, coal is economical only for heat and power; with oil at $35/barrel, it is also economical for production of ammonia and methanol; and with oil at $40/barrel, it is economical even for the production of ethylene and gasoline. Vegetable products are likewise viable alternatives as raw materials for ethylene and gasoline production as oil prices reach $50/barrel.[3]

Studies at the Agriculture Department's Northern Regional Research Center in Peoria, Illinois have identified 34 plants, including milkweed, quackgrass, poinsettia, and soro thistle, as potential sources for the hydrocarbons necessary to manufacture rubber and plastics. Some produce rubber, some oil, and some both; some even have sufficient remaining protein and carbohydrates for cattle feed or for pulp for papermaking. [4]

FIGURE 16-2.   Contemporary oil refinery. (*Photo courtesy DOE.*)

It is possible that, in addition, cellulose can be shredded, milled, and then fermented to sugars. This process can be used for almost all sources of cellulose—newspaper, cotton, and rice hulls. It may, therefore, be a useful technique for recovery of various cellulose-based wastes.[5]

For the near future, however, oil and natural gas will continue to be the major raw materials for most organic chemicals. In the United States, in 1975, about 11% of the ~1100 gallons (4160 L) of petroleum consumed per person annually was used for petrochemicals. This amounts to a total U.S. commitment of 660 million barrels annually for petrochemicals.[6]

When considering the organic chemicals industry, it is necessary to consider three types of activity:

1. Petroleum companies, which primarily convert crude into refined stocks to be used for fuels with high-performance characteristics or as raw materials for the production of other chemicals.
2. The basic chemicals industry, which produces bulk chemicals, usually pure, single compounds.
3. The use of the bulk chemicals as raw materials for pharmaceuticals, synthetic fibers, pesticides, solvents, detergents, fertilizers, coatings, construction materials, and so forth.

Sometimes one company will do all three, or at least two, of these functions. For example, of the 50 largest bulk chemical companies, 13 are actually primarily in the petroleum field. [7]

Separation and Conversion Processes

Petroleum deposits are located all over the world. There is no exact method used, however, to locate the precise spot to drill. Modern exploration methods include magnetic, gravity, and seismic (shock) measurements, but even with these techniques there is no sure way to determine whether or not oil is present except to drill.

The first step after receipt of the raw petroleum is refining. The complex mixture of the various petroleum products is initially separated on the basis of the relative volatilities of the components. This process is essentially a large-scale fractional distillation: as the temperature is gradually increased, the various components are vaporized in order of decreasing vapor pressures.

The dissolved gases, methane (CH_4), the propanes, and the butanes, vaporize first. The latter two will liquify under pressure and are used as bottled gas: the former is sold as natural gas. Or these substances can form the basis of other chemical derivatives.

Petroleum ether, a mixture of the most volatile of the liquid hydrocarbons, is next separated.

Gasoline distills by the time the temperature reaches 50-200°C. It is also a complex mixture of hydrocarbons, but primarily those containing 6-12 carbon atoms/molecule.

After the gasoline, kerosene, fuel oil, and diesel oil are vaporized. Then, at reduced pressure, the low-volatility, high molecular weight waxes and lubricants distill. These substances would not distill at ordinary pressures, for the required temperature would be so great that they would instead decompose.

Left are the nondistillable asphalt-type residues of a very high molecular weight, and also the tarry materials that have formed during heating.

One problem, long experienced by the petroleum industry, is the uneven demand for the various products—uneven not only from year to year, but also seasonally. Until the great popularity of the automobile, kerosene was much more desirable than gasoline. Now, of course, the reverse is true. In the winter, the heating fuels are more in demand; in the summer, gasoline consumption rises. Regardless of which of the substances is in more demand, further refining is always needed to convert a less valuable substance into a more valuable one. Now, much effort is devoted to the conversion of petroleum ether and kerosene to gasoline, which comprises about 50% of the total product.

The larger molecules, those containing 12-20 carbon atoms/molecule, are then split into smaller molecules with a volatility suitable for gasoline, by a procedure known as "cracking." In addition to splitting apart larger molecules, cracking also produces large quantities of ethylene and other unsaturated hydrocarbons (those with double and/or triple bonds) by removing hydrogen from a saturated hydrocarbon. For example:

$$\begin{array}{c} CH_3 \\ \diagdown \\ \diagup \quad CH-CH_3 \\ CH_3 \end{array} \longrightarrow \begin{array}{c} CH_3 \\ \diagdown \\ \diagup \quad C=CH_2 + H_2 \\ CH_3 \end{array}$$

Isobutane   Isobutene

or

Isobutane $\longrightarrow$ $CH_3-CH=CH_2 + CH_4$

Propene

The heavier molecules present even more possibilities as to potential products. The net effect thus is to produce small hydrocarbons from large ones and to simultaneously increase the amount of unsaturation.

Some of this cracking is accomplished in the initial heating. Most, though, is done in the presence of solid catalysts, usually clay derivatives such as silicas. Many alkenes (isobutene, propene,...) are formed. The less volatile of these are particularly important, for not only do they enhance the antiknock properties of gasoline, but they also can be separated out, and can be used as intermediates for the production of other chemicals. Of these, ethylene (C_2H_4) is the most valuable. It is actually the most heavily used petrochemical, and forms the basis for the production of polyethylene, for packing materials; ethylene oxide (CH_2-CH_2), used in epoxy resins; ethanol (CH_3CH_2OH); ethylene glycol, an

$$\begin{array}{c} CH_2-CH_2 \\ \diagdown \quad \diagup \\ :O: \end{array}$$

antifreeze; and butadiene, widely used in the manufacture of synthetic rubber.

The next step is the formation of aromatic compounds from cyclic and straight chain hydrocarbons. This is done at high temperatures, in the presence of catalysts. An example would be the conversion of methyl cyclohexane to toluene:

$$\begin{array}{ccc} CH_3 & -3H_2 & CH_3 \\ \bigcirc & \xrightarrow{\quad catalyst \quad} & \bigcirc \end{array}$$

Toluene is very beneficial in gasoline, for it increases the octane number without the use of lead.

The aromatics can then be converted to specific pharmaceuticals, pesticides, or synthetic fibers.

Pollution in the Petroleum Processing Industry

When discussing potential pollution problems, it is desirable to discuss first those of the petroleum companies, then to proceed to the bulk chemicals industry, and, lastly, after discussion of some particular industrial processes that are used to produce specific chemicals, the problems encountered in a typical chemical company. Some problems (and solutions) are common to the latter two.

Drilling and Transportation Effects

The potential pollution problems start with the drilling. A typical petroleum drilling rig is shown in Figure 16-3. The first oil in the United States was found at a depth of only 21 m in Titusville, Pennsylvania. Today, it is often necessary to drill very deep wells, requiring thousands of feet of well casing, tubing, and drill pipe.

The ordinary oil recovery methods extract only about one-third of the oil potentially available at a specific site. Because of concerns about energy availabil-

FIGURE 16-3.   A typical oil-pumping rig.

ity, and the projected shortages of oil in the future, enhanced oil recovery techniques have been developed. About 40% of current U.S. oil production is by the "water-flood" method: water is pumped into an oil reservoir under pressure. This water sweeps a large portion of the remaining oil toward the producing wells. Other methods, using detergents, solvents, acids, and/or heat, have been studied, but are not yet at the commercial stage.

The extraction of oil from oil shale would create, perhaps, the greatest technical and environmental difficulties. This procedure is still in the developmental stage. Interest slowed in the mid-1980s, due to the availability of relatively plentiful and inexpensive imported oil. On the average, it is necessary to process 1 ton of shale for every 25 gallons of oil produced. This generates a significant solid waste ("tailings") problem (Figure 16-4). Current technology also requires a large amount of water. This water is very difficult to treat, and cannot currently be recycled because of being chemically bound with the solid wastes.

Some of the world's largest oil deposits are located offshore (Figure 16-5); in the United States, they are located near states such as California, Louisiana, Texas, and Florida. Leakage and accidents involving offshore drilling operations could be

FIGURE 16-4.   Tailings generated by oil shale extraction. (*Photo courtesy DOE.*)

FIGURE 16-5.   Offshore oil drilling rig. (*Photo courtesy DOE.*)

severe problems, for the oil can devastate marine life. For example, the oil penetrates into the feathers of aquatic birds, making flight impossible due to the added weight. The oil, lighter than water, floats on top, forming a film. This changes the rate of evaporation of that water, as well as severely affecting the many marine plants that also concentrate at the surface. In addition, oil is a good solvent for other substances, many of which are toxic. One good example is the pesticide DDT, which is 10^8 times more soluble in oil than in water.

The long-term effects of oil spills are still unclear, however. A 1975 study[8] of the many oil-laden German and U.S. ships sunk during World War II has tentatively led to the hope that the impacts are rapidly minimized by natural forces. It is likely that large spills may not be, in the long run, as detrimental as widespread constant, small discharges.

Actually, it does not appear to be the offshore drilling operations that are currently the major danger, but, instead, the oil tankers and supertankers (Figure

16-6). Of the 20,000 offshore wells drilled since 1947, only two have resulted in a blowout with a spill that reached shore. Until the very large, long term, and well-publicized blowout of the Mexican offshore drilling rig at Ciudad del Carmen on June 3, 1979 (the oil of which reached the ecologically sensitive Texas coast), there hadn't been any large oil spills due to offshore drilling since 1972. During that same time period, there had been many very large, well-reported sinkings of supertankers off the coast of the United States and other countries—ships such as the Amoco Cadiz, which dumped about 2 million barrels (84 million gallons) of oil into French coastal waters in 1978, and, more recently, the Exxon Valdez, which leaked 11 million gallons into Alaska's Prince William Sound in 1989.

During January 1991, another cause for a major oil spill (and air pollution) became apparent: terrorism and/or war. Iraq intentionally poured oil from on-shore Kuwaiti storage tanks into the Persian Gulf through deep-water pipelines used to load supertankers, and set 550 oil wells on fire. Estimates are that 6 to 8 million barrels (250-330 million gallons) had poured into the Gulf before U.S. aircraft were able to stop the flow by bombs.

Methods other than supertankers also are used for transportation of petroleum. These include barges, pipelines, railroad tank cars, and highway tank trucks. Large, towed barges are used extensively in the Great Lakes and in the approximately

FIGURE 16-6.   Supertanker used for liquefied natural gas transport. (*Photo courtesy DOE.*)

30,000 miles of inland waterways for transportation of petroleum inland. The majority of the remainder of the transportation, particularly of petroleum fuels, is by highway tank truck.

Whenever a large spill does occur on a waterway, it must be contained as rapidly as possible. Spills of oil on water are typically treated by "booms"[9,10] (Figure 16-7). The spill is generally surrounded by a sorbent material (encased in a fabric to form a sausage-like shape), that does not allow the liquid to penetrate. Since oil floats on water, booms must, therefore, also float on the water surface, both before and after sorption has occurred.

Most commonly, booms are made from one of the polyolefins, particularly the melt-blown, carded products. This material is thick and highly porous, with minimal strength. The sorbent should be encased in, for example, a thinner spun and needled polyoefin fabric. The hydrocarbons are not very tightly bound to the polyolefin surfaces; hence, the booms can be run through wringers for subsequent oil recovery.

The oil spill problem is extremely difficult because oil can spread on water to 0.01 in thick, over a 25-square-mile area, in 8 hours.[11] Much research is being done

FIGURE 16-7.   Booms such as these are almost universally used to contain oil spills as the first stage in the cleanup operation. (*Photo courtesy JV. Manufacturing Co, Inc.*)

to develop new methods and approaches. Using surface-film-forming chemicals to minimize the oil spreading, or to even drive the oil back into a thicker layer, is one possibility.

Once a spill is contained, the petroleum must then be collected. A number of different products and techniques have been developed to accomplish this. These include mechanical skimming and microbes to consume and degrade the oil (replicating what occurs in nature). Creation of a vortex by an impeller, such as that generated by stirring a glass of water, can draw the oil slick into the center where it can be removed by pumping.[12,13]

Another sorbent that has shown considerable promise for both containment and cleanup of oil spills is a mixture comprised of cotton lint and particulate wood fiber. Tests have indicated that this very light material (13 lb/ft^3) will not only contain and collect oils from water surfaces when encased as booms, but that, in the bulk form, it can be used to clean oil-contaminated shoreline. If the sorbent is spread on oil-soaked rocks and permitted to remain there about 3 hours, the saturated product can then be hosed off into the water. After recovery with a filter screen, the oil laden sorbent can be burned for fuel. In contrast to the synthetic polymers such as the ployolefins, the sorbent is biodegradable and totally nontoxic, even if ingested by ducks, fish, or other aquatic animals.[14]

The gelation of crude oil, to convert it from a liquid to a solid within a distressed tanker, is also being studied. The liquid organic gelling agents would be dissolved in the oil to form a gelled compound that physically entraps the oil. If this mixture escaped from a ship, it would float in a big mass.

Another approach, the use of bacteria for shoreline bioremediation, was first tested for oil spill cleanup in the March 1989 spill of the Exxon Valdez. This process utilizes naturally occurring bacteria, stimulated with a garden variety, water-soluble fertilizer. In this test, the organisms fed on the oil that had seeped almost 2 ft into the gravel and sand of the Prince William Sound. After only 3 weeks, the fertilized areas were nearly free of oil for 12 inches, whereas the control areas were still oil-coated.[15]

The Coast Guard has significant responsibility in spill control, dependent upon whether the spill is inside or outside the U.S. territorial limits. Outside the territorial limits, the U.S. Geological Authority has responsibility; inside, the Coast Guard does, and it can fine for oil discharges with penalties up to $5000. The Coast Guard has the authority to oversee private oil-spill cleanup activities, or, if the response is not sufficient, to personally clean up the spill and bill the polluter. Three teams of specially trained and equipped personnel are available, one on each of the Atlantic, Gulf, and Pacific Coasts. In addition to helping to clean up spills, it also has an extensive remote sensing program to detect spills. The Coast Guard is developing methods to determine the source of a spill, so violators can be prosecuted (techniques include use of infrared and fluorescence spectroscopy, gas chromatography, and thin layer chromatography).[16]

The use of pipelines is the cheapest transportation method—the most famous pipeline currently being the Trans-Alaska pipeline (Figure 16-8), designed to bring oil from the North Slope of Alaska to a tanker base in an all-weather bay (Valdez) located in the southern part of Alaska. This remarkable engineering feat crossed 800 miles of underdeveloped land, rich in wildlife and scenery.

The potential environmental hazards of the pipeline include:

1. Leaks and breaks.
2. The movement of relatively warm oil in the pipeline, as well as the additional frictional heat generated by the heavy crude, could melt the permafrost, potentially leading to collapse of pipeline supports, resulting in more leaks and breaks.
3. The very slow return of all vegetation due to the short growing seasons.
4. Disruption of wildlife migratory paths, in spite of bridges, where the pipe is raised, to permit animals to travel underneath.

The environmental impact of this pipeline appears to have been minimal, but it must be continually and carefully monitored, because the problems created by a spill could be very serious for the state due to the fragile environment.

FIGURE 16-8.

Treatment Methods at the Refineries

Most of the wastes generated at the petroleum refineries create water pollution problems. A few steps do create air emissions, but, typically, these vapors and/or particulates are collected by wet scrubbers, and thus degenerate to a water treatment problem. Generally, 3–8 gallons of wastewater are produced per barrel of oil.

Petroleum refineries usually employ multistage waste treatment procedures, with separate collection of the various types of waste. In general, the refineries try to 1) regenerate the waste to form products low in sulfur, 2) recover as much as possible of the various solvents for reuse, and 3) scrub any process gases.

The Standard Oil Complex—Texas City, Texas

Figure 16-9 illustrates a water treatment facility installed at the Standard Oil Complex in Texas City, Texas. This particular facility is used to purify, not only refinery water, but also the chemical plant process water, storm water, and the water used as oil-tanker ballast. Most of the solids and oil are recovered in the primary stage by separators, skimmers, or filters; the usable oil is recycled to the process, and the sludges are used as soil conditioners. Air is then pumped into the water, chemical coagulant added, and the remaining oil and suspended solids filtered out. These solids are again used as soil conditioners. The soluble contaminants are then consumed by microorganisms in an activated sludge-type process. These microorganisms settle, and are mainly recycled. The water is then filtered before it is released into Galveston Bay.

The distinction between this process and a conventional municipal one is that the bacteria are used to remove only the dissolved contaminants, not the suspended ones. This saves money, for there are fewer new bacteria that must then be disposed of. The net result is an effective system that economically removes nearly all of the

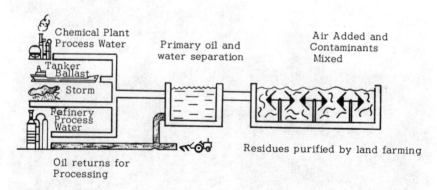

FIGURE 16-9.   Schematic of the new water treatment facility at the Standard Oil complex in Texas City, Texas.

suspended matter and dissolved contaminants from 23 million gallons of water daily, equivalent to that generated by a city of about 300,000.[17]

Byproduct Recovery
There are some byproducts from the refinery operations, and some recycling, which can be economically advantageous. These include the following.[19]

1. Sulfuric acid can be recovered from the sludges that are produced by the treating of various of the oils with sulfuric acid, in order to de-emulsify them for separation from the wastewater.
2. Sulfuric acid, used to catalyze the alkylation of olefins and alkanes for substances such as leaded gasoline, and other spent acids used in the acid treatment of oils and waxes can be recycled and reused.
3. Catalytically cracking the naphtha fractions produces phenol-containing bases that can be recovered and sold.
4. The phenols can also be used as refinery fuels.
5. Recovered waste acids can be used for oil treatment within the plant.
6. Aluminum (which is a cracking catalyst) can be recovered from the hydrocarbon sludges for further use for solids separation, phosphate removal, and sludge conditioning.
7. Acid oils can be recovered from the waste caustics by reacting the latter with acids.
8. Sludge from the treatment of the boiler water can be used to neutralize some of the wastewater.
9. Treated water can supplement the normal refinery water.
10. Ammonia and hydrogen sulfide can be recovered from process water, and eventually converted to ammonium sulfate for fertilizer.
11. Small, single-cell proteins can be produced from petroleum and petroleum

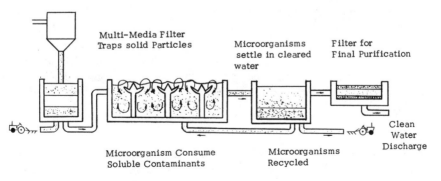

FIGURE 16-9.   *(Continued)*

wastes by letting microorganisms grow by feeding on hydrocarbons. These can potentially form animal or human food supplements.

Chemical Process Industries

The chemical process industries are concerned with taking the petroleum feed stocks, primarily the unsaturated alkenes and aromatics, and forming from them other intermediates or finished products. For example, they produce polymer resins that are then sold to a textile manufacturer; or they produce plastics, which are often sold directly in retail markets. Two of the major customers for these industries are producers of pesticides and detergents (though much also goes for the production of paints and lacquers) and producers of medicinal products (including antibiotics, tranquilizers, analgesics, hormones, and vitamins). The largest general market of all, however, is for the production of various polymers.

The Polymer Industry

A large number of various types of polymer products are manufactured on a large scale. These include plastics, synthetic fibers, and rubber.

Polymers are large molecules, made up of many repeated smaller (monomer) units. One of the most important polymers, cellulose, occurs naturally. Cellulose is composed of many repeating glucose units. Many of the important polymers, however, are man-made.

Polymerization can occur in two ways:

1. Addition polymerization, which occurs when the molecules simply add together to form giant molecules; and
2. Condensation polymerization, which occurs when small molecules are ejected as giant molecules build up.

Polymers can be classified in several ways. One way is according to their structural rigidity. Thermosetting polymers are very rigid; thermoplastic polymers are mobile at high temperatures; elastomers are the synthetic rubbers. Polymers can also be classified according to their monomeric origin. Polyolefins are made from olefin monomers, or units containing double bonds. An example is polyethylene, made from $H_2C=CH_2$ monomers and used as a packaging and structural material. The polyethylene structure is $-H_2C-CH_2-CH_2-CH_2-$. Polyamides

are composed of amide monomers $\left(\begin{array}{c} O \\ \parallel \\ R-N-C- \\ | \\ H \end{array} \right)$. Nylon, with a partial structural

$$\overset{\text{O}}{\overset{\|}{}}\qquad\overset{\text{O}}{\overset{\|}{}}$$

formula of $-(CH_2)_5-C-NH-(CH_2)_5-C-NH-(CH_2)_5-$, is a polyamide.

Polyesters, made from ester monomers $\left(-\overset{\overset{\text{O}}{\|}}{C}-O-R \right)$, might have structural

formulas such as $\left(-C_2H_4-O-\overset{\overset{\text{O}}{\|}}{C}-CH_2- \right)$. Both the nylons and the polyesters
are widely used as fabrics.

Plastics

A plastic is, by definition, any substance that can be molded and shaped. In practice, many substances that meet this definition (synthetic fabrics, synthetic rubber, etc.) are not truly considered plastics. On the other hand, one of the first major plastics, bakelite, is really very hard and rigid under ordinary conditions. Only during its manufacture is it soft and pliable.

Many different plastics are produced, with many different properties. We shall consider only the main ones.

Polyethylene is the plastic produced in the largest volume. As mentioned earlier, it is manufactured by polymerization of ethylene molecules:

The formation of low-density polyethylene is initiated by either heating ethylene at high pressures (15,000-50,000 psi) and/or by the use of "initiators" such as organic peroxides. Peroxides, with an R-O-O-R structure, can break apart to form two very reactive free radicals, $R-O\cdot + \cdot O-R$, each distinguished by an unpaired electron. Each free radical can combine with an ethylene molecule to form a new free radical, one unit ($-CH_2CH_2-$) greater than the original:

Chain propagation continues until the free radical is destroyed by 1) combining with another free radical, or 2) reacting with an inhibitor. The result is a polymer product containing giant molecules of different molecular weights. Even a small amount of initiator can produce a great deal of polymerization. The R—O—end groups are insignificant: the molecules are essentially aggregates of C_2H_4 units added together.

High-density polyethylene is, instead, catalyzed by metallic halides and oxides or aluminum alkyls. Less high-density polyethylene is produced annually than the low-density product.

Polyethylene is used in the manufacture of plastic bags, toys, films, sheets, and bottles.

Polypropylene is a harder, tougher, more durable plastic, formed by the addition of propylene monomers:

$$
\begin{array}{ccc}
\text{H} & \text{H} & \text{H} \\
\backslash & | & | \\
\text{C}=\text{C}-\text{C}-\text{H} & \xrightarrow{\text{polymerization}} & -\text{C}-\text{C}-\text{C}-\text{C}-\text{C}- \;. \\
/ & | & \\
\text{H} & \text{H} &
\end{array}
$$

The added $-CH_3$ groups make it a more rigid substance. Polypropylene is used in the manufacture of indoor-outdoor carpets, pipes, valves, and bottles.

Polyvinyl chloride (PVC) if formed by the polymerization of vinyl chloride monomers:

$$
\begin{array}{ccc}
\text{H} & & \text{H} \\
\backslash & & / \\
& \text{C}=\text{C} & \xrightarrow{\text{polymerization}} \quad -\text{C}-\text{C}-\text{C}-\text{C}- \;. \\
/ & & \backslash \\
\text{H} & & \text{Cl}
\end{array}
$$

PVC is a very common constituent of rigid plastic piping and many other materials such as films and sheet, flooring, wire insulation, automotive parts, and adhesives and coatings. In the early 1970s, PVC plastics were linked to the development of a rare form of liver cancer in workers. Major changes in the PVC industry have resulted. Polyvinyl chloride is produced by free radical polymerization of vinyl chloride using an initiator such as a peroxide, in a manner similar to the production of polyethylene.

Bakelite, one of the first plastics to be produced, is a phenolic resin; that is, it is formed by the addition of formaldehyde and phenol molecules:

Phenol   Formaldehyde   Bakelite intermediate

Bakelite

The multitudinous $-CH_2-$ bridges lead to a very hard, rigid material.

Synthetic Fibers

The first of the synthetic fibers, made in 1891, was rayon. Rayon is produced by treating cellulose with sodium hydroxide (NaOH) and carbon disulfide (CS_2). This process breaks up the short cellulose polymers, which are then reformed into fine, continuous filaments in a device called a spinnaret. These filaments are then spun into threads of the desired size, which can be woven into fabrics.

Nylon, one of the most successful of the synthetic fibers, was developed in 1935. Its production is by "condensation." This type of polymerization occurs by the combination of (usually two types of) small molecules, accompanied by elimination of simpler species such as water.

For this type of process to be feasible, two conditions must be satisfied:

1. One end of Molecule A must be able to interact with an end of B so as to eject a small molecule.
2. Each of the two types of molecules must contain two similar functional groups

so that, after they combine, they still will have free ends that can continue to react to further extend the polymer.

Nylon is formed by the reaction of diaminohexane and adipic acid:

$$
\begin{array}{c}
\text{H} \quad \text{H H H H H H} \quad \text{H} \\
\diagdown \quad | \ | \ | \ | \ | \ | \quad \diagup \\
\text{N—C—C—C—C—C—C—N} \\
\diagup \quad | \ | \ | \ | \ | \ | \quad \diagdown \\
\text{H} \quad \text{H H H H H H} \quad \text{H}
\end{array}
\quad + \quad
\begin{array}{c}
\text{O} \quad \text{H H H H} \quad \text{O} \\
\diagdown\diagdown \quad | \ | \ | \ | \quad \diagup\diagup \\
\text{C—C—C—C—C—C} \\
\diagup \quad | \ | \ | \ | \quad \diagdown \\
\text{HO} \quad \text{H H H H} \quad \text{OH}
\end{array}
\longrightarrow
$$

Diaminohexane   Adipic acid

$$
\begin{array}{c}
\text{H} \quad \text{H H H H H H} \quad \text{O H H H H} \quad \text{O} \\
\diagdown \quad | \ | \ | \ | \ | \ | \quad \| \ | \ | \ | \ | \quad \diagup\diagup \\
\text{N—C—C—C—C—C—C—N—C—C—C—C—C—C} \\
\diagup \quad | \ | \ | \ | \ | \ | \quad \ | \ | \ | \ | \quad \diagdown \\
\text{H} \quad \text{H H H H H H} \quad \text{H H H H} \quad \text{OH}
\end{array}
\quad + \ \text{H}_2\text{O}.
$$

Nylon

Both ends of the new nylon molecule are thus available to continue the polymerization.

The nylon filaments are then produced by pressing hot, melted nylon through the holes in a spinnaret. The filament thickness can vary from extremely fine, as used in nylon pantyhose, to the very large, durable fibers used in truck tires. Specialized nylons have also been developed. These include Nomex, fairly high-temperature-resistant, used in fabric filters in baghouses.

Synthetic Rubber
Natural rubber is a polymer of isoprene,

$$
\begin{array}{c}
\text{H} \\
| \\
\text{H} \quad \text{H—C—H} \quad \text{H} \\
\diagdown \quad | \quad \text{H} \quad \diagup \\
\text{C} = \text{C—C} = \text{C} \\
\diagup \quad \quad \quad \diagdown \\
\text{H} \quad \quad \quad \text{H}
\end{array}
$$

Isoprene

All rubber formerly was manufactured from latex, a sap from rubber trees that grow only in the South Pacific. When Malaya and Indonesia fell to the Japanese in 1942, the U.S. supply of rubber was nonexistent. Very rapidly, synthetic rubber plants were built, employing a German process, and the U.S. production of synthetic rubber increased from zero, in 1940, to 670,000 tons, in 1974.[20]

The first synthetic rubber was produced by the polymerization of 70 parts butadiene and 30 parts styrene:

Styrene   Butadiene

Synthetic rubber

Today, rubber is also manufactured from monomers such as polybutadiene, polyisoprene, and neoprene. The addition of carbon black has improved the properties of the synthetic rubbers by reinforcing the elasticity, stiffness, and strength.

Environmental Problems of the Organic Chemicals Industry

The potential environmental problems related to organic chemicals production fall into several classes:

1. The effluent from organic chemicals plants is generally rich in BOD. This is really the most serious problem, but it arises primarily for economic reasons. Even if it is possible to produce byproducts, usually they are in little demand. It tends to be expensive to further treat the effluent, to burn the residual organics to CO_2 and H_2O, or to properly haul away and dispose of the high-BOD wastes. Many of the problems arise because of poor plant operation. Leakage of BOD due to faulty valves, spilled chemicals, and improperly trapped suspended materials are not uncommon. Older chemical plants, as in all industries, are less likely to employ state-of-the-art pollution control equipment.
2. There are, in addition, many hazardous chemicals that can cause problems if they are released to the environment (Figures 16-10 to 16-12). One example,

FIGURE 16-10. Chemical dumps frequently are disaster areas in terms of the amount of chemicals released directly to the environment. (*Photo courtesy Wisconsin DNR.*)

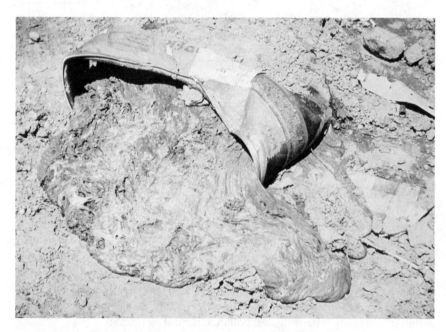

FIGURE 16-11. Containers frequently become crushed, spilling their contents. (*Photo courtesy Wisconsin DNR.*)

FIGURE 16-12.   Barrels can open up, disgorging their contents. (*Photo courtesy Wisconsin DNR.*)

discussed in detail in Chapter 1, is the polychlorinated biphenyls (PCBs), which were long used as plasticizers in paints, resins, and plastics, and as heat transfer fluids and insulators in (for example) capacitors. These persist in nature, and accumulate in organisms, especially in the fatty tissues. They have been largely prohibited in the United States Phthalate esters, such as Di-2-ethylhexylphthal-ate, are also used as plasticizers in many polymer substances. They can help maintain a flexible structure by preventing crystallization of the polymer. Unfortunately, there have been instances where they have leached out of plastic bottles and plastic tubing, contaminating the stored substances, including blood plasma.

$$\begin{array}{c}
\text{O} \qquad\qquad \text{CH}_2\,\text{CH}_3 \\
\parallel \qquad\qquad\quad | \\
\text{C}-\text{O}-\text{CH}_2-\text{CH}-\text{CH}_2\,\text{CH}_2\,\text{CH}_2\,\text{CH}_3, \\
\\
\text{C}-\text{O}-\text{CH}_2-\text{CH}-\text{CH}_2\,\text{CH}_2\,\text{CH}_2\,\text{CH}_3 \\
\parallel \qquad\qquad\quad | \\
\text{O} \qquad\qquad \text{CH}_2\,\text{CH}_3
\end{array}$$

3. The air pollution associate... ith chemical plants is primarily particulates, sulfur dioxide (SO_2), and various hydrocarbons. These can be problems within the plant as well as in the outside environment (Figure 16-13). Substances such as ethylene and other olefins inhibit plant growth, as does SO_2.

FIGURE 16-13.   Fume collection ducts crisscross the ceiling of the Pearsall Chemical Corp. plant in La Porte, Texas. (*Photo courtesy Pearsall Chemical Corp.*)

4. There is also the problem of simply disposing of the products. Many of the polymer substances—the plastics used for packaging, the synthetic rubber used in tires—constitute a major solid waste problem, for they are not biodegradable. Polyvinyl chloride, when incinerated, can rapidly form hydrochloric acid (HCl). Effort is being made in the development of new types of plastics, ones that are biodegradable, but these products are also controversial.

Toxic Substances

Many organic substances are known to be, or at least suspected of being, carcinogenic. One well-known example is the chemical Kepone, a chlorinated hydrocarbon pesticide produced by Allied Chemical that seriously affected many exposed workers. After being banned, even its disposal was for a long time under debate. For a while, the Kepone had been discharged into the James River. Allied was fined $13.2 million (plus the costs of cleaning up the river) for this illegal discharge and for the expenses incurred by the EPA in sampling, monitoring, and analyzing various substances, an additional $1 million plus. The remaining pesticide had been stored in silos near Boise, Idaho; in addition, there was Kepone-contaminated sludge at the Hopewell, Virginia sewage treatment plant; Kepone-contaminated soil around the Life Science Products Company plant, which had manufactured the chemical for Allied; and an additional 85,000 lb stored in scrap metal in Portsmouth, Virginia and Baltimore, Maryland. Allied had proposed to entomb the Kepone—some of which was considered unburnable—in abandoned missile silos near Boise, Idaho. This was turned down by Idaho.[21] Tests in March 1977 finally showed that the Kepone could be safely incinerated at high temperatures. At about 2000°F, the pesticide is broken down to CO_2, H_2O vapor, and NaCl.[22]

Another toxic substance is vinyl chloride ($CH_2 = CHCl$), or chloroethene, manufactured from either ethylene or acetylene and chlorine. The most common procedure is by conversion of ethylene to ethylene dichloride ($ClCH_2CH_2Cl$), which is then thermally cracked to $CH_2 = CHCl$ and HCl. The HCl is then reacted with ethylene to form more ethylene dichloride, and the cycle continues.

Antiosarcoma, a rare liver cancer, has been linked to exposure to vinyl chloride; ethylene dichloride is also a suspected carcinogen. It is thus essential that contact with these substances be minimized.

The major source of vinyl chloride waste in a plant is in the process air stream. This gas typically contains other chlorinated hydrocarbons as well. One of the best solutions to this problem is to incinerate the waste gas, followed by wet, caustic scrubbing of the flue gas from the incinerator. The burning of chlorinated hydrocarbons results in HCl as well as CO_2 and H_2O. This acidifies the scrubber liquor, which can be neutralized by a caustic, often at prohibitive costs. The difficulties associated with liquor disposal depend upon the location. Discharge of a weakly acidic or salt solution into the ocean is not a problem, but it can be in other locations.

The second largest source of vinyl chloride emissions is in the loading operations—loading into tank cars, tank trucks, barges, or ships. Vinyl chloride is a relatively volatile liquid, and thus vapors are always a serious problem. Solutions involve steps such as purging the system and equipment before and after loading with an inert gas.

In addition, precautions taken at the plant include checking seals, valves, and proper decommissioning of equipment for inspection or maintenance.[23]

A number of "biorefractory" chemicals are causing tastes and odors in drinking water; some of these chemicals also taint seafood. The lower end of the Mississippi River, the source of water for half of Louisiana's population, has been found to contain 34 different chemicals. These substances are all low molecular weight chlorinated aliphatic (straight chain) and aromatic hydrocarbons with a low volatility that are not amenable to biological treatment. Included are vinyl chloride, chloroform, and perchloroethylene.

What can be done? Five techniques of waste treatment have been looked at in recent years. These are ozonation, carbon adsorption, solvent extraction, anaerobic/aerobic treatment, and air stripping.[24] It is hoped that one or more of these will be effective in the near future.

Waste Control and Recovery

The first step for waste chemicals control is collection within the plant; the next is proper disposal. What methods are used to control these potential pollutants? The organic chemicals industry is complex due to the number of products and processes utilized; the treatment processes are thus also complex and often very specialized. All plants do attempt to recover byproducts and reduce their waste loads accordingly. If the recovered substance does not currently have a market, it may be used as a fuel. Hydrogen is commonly used in that manner.

A few examples of possible recovery processes will be considered here.

Ethylene: Ethylene is the organic chemical produced in largest volume. In the past, most ethylene had been made from fairly light feedstocks, such as ethane removed from natural gas. Currently there is a switchover to instead cracking oil-derived heavy liquid feedstocks such as naphtha and gas oil. Ethylene can also be made directly from crude oil. Ethylene production generates approximately 15 gallons of waste water/ton of product. This ethylene waste water is about 2.5% NaOH, 1% Na_2S (primarily from the waste water treatment already undergone), and 6.6% phenols (from the cracking). This condensate is "stripped" by steam to remove all of the nonphenols and perhaps 20–25% of the phenol. The remainder of the phenol can be removed by contacting the condensate with fresh feed. This phenol-rich feed is then taken to the cracking furnace. The cleaned water can be reused in the plant. The NaOH and Na_2S are not recovered, though the Na_2S is usually oxidized to reduce the BOD.

Polyethylene: Most plants that produce ethylene also produce polyethylene. The major pollution problems associated with polyethylene are the spilled "fluffs" and "pellets." Not only are they unsightly in the effluent, but they also slow down the rate of oxygen transfer. A surface drainage separator can be used to separate out the polyethylene (which floats) from the remainder of the process water. This is usually done by installing a basin with a baffle, extending 3 ft below the surface. The trapped polyethylene is usually disposed of by landfill.

Hydrogen: Hydrogen is a common product from various refinery operations, hydrotreating, and hydrocracking, which produce multitudinous organic compounds. This hydrogen can typically be recovered from the gaseous effluent by:

1. Wet scrubbing with water solutions of carbonates or ethanolamines;
2. Absorption onto activated carbon or other "molecular sieves";
3. Cryogenic scrubbing, where the higher boiling impurities can be condensed out; and
4. Allowing hydrogen to be selectively passed through a palladium membrane.

Other hydrocarbons can typically be recovered by processes that include distillation, crystallization, absorption, extraction, ion exchange, and oxidation.[25]

Rather than depend on end-of-pipe treatments, many chemical companies have emphasized waste reduction, especially by internal plant recycling. A 1988 survey by the Chemical Manufacturers Association[26] of 522 chemical plants found the following trends, compared to 1981:

1. A decrease in solid waste generated by 41.5%;
2. 18.6% decline in wastewater generation;
3. Decrease in landfill disposal by 13.6%;
4. An increase in incineration of 50.2%;
5. An increase in the use of injection wells from 5.6 to 6.5%;
6. A decrease in use of publicly owned treatment plants from 9.8% to 4.2%.

The plants surveyed produced a total of 218.8 million tons of hazardous waste in 1987. Of this, 211.6 million tons was wastewater and 7.2 million tons was solids. Figure 16-14 summarizes the disposition of these hazardous wastes.

A number of approaches have been taken to control air pollution. Some of these are summarized in Table 16-1.

Disposal of Toxic Substances

Even when toxic wastes are properly collected, the plant is still faced with the question of final disposal. Recently, there have been many horror stories of

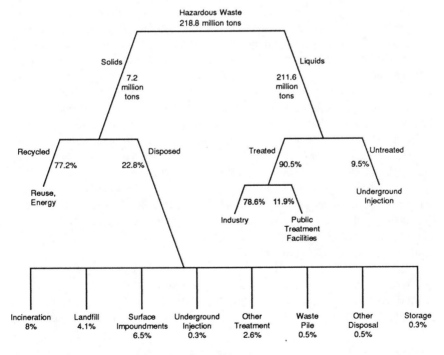

FIGURE 16-14.   Summary of the disposition of hazardous wastes.

TABLE 16-1.   Air Pollution Control Strategies[a]

Method	Organic Vapors (VOC)	Inorganic Vapors	Particulates	SO$_x$ & NO$_x$
Incineration	X			
Adsorption	X			
Absorption	X	X		
Condensation	X			
Baghouses			X	X
Precipitators			X	X
Wet Scrubbers		X	X	X
Chemical Reaction				X

[a]Data from "Wiping out Air Pollution," *Chem. Eng.* **97**, *No. 9,* Sept 1990, p. 106.

hazardous wastes returning to haunt not only the industry, but the citizens who are unfortunate enough to live near an abandoned, inadequate, or illegal dump site. These dumps are located all over the United States. Some of the more obvious of these cases, recently reported nationally, are located in places such as Niagara Falls, New York (Love Canal); Lowell, Massachusetts; Iberville Parish, Louisiana; Montague, Michigan; Houston, Texas; Philadelphia, Pennsylvania; West Point, Kentucky; and Elizabeth, New Jersey,[27-31] Few locations are immune.

The Resource Conservation and Recovery Act gives the EPA authority over only current or future dump sites. Many of these abandoned or inactive sites can be cleaned (by EPA mandate) only if 1) the site is an imminent or substantial hazard to human health or the environment, and 2) if the owner has sufficient funds. In many cases, unfortunately, the owner has died, gone bankrupt, or is unknown.

A few industries do still carry on "midnight dumping"—dumping wastes along the side of the road when no one is looking, for example. This is particularly true in the industrialized Northeast, where licensed, secure disposal sites are few, and those are operating at full capacity. There are signs that even organized crime has entered the business of illegal hazardous waste disposal.[34]

Many industries, on the other hand, are disposing of their wastes properly. Manufacturers such as 3M Corporation, Eastman Kodak Company, and Dow Chemical Company use high-temperature incineration techniques.[35] Numerous others are encapsulating their wastes and then disposing of them in properly secured landfills, as prescribed (Figure 16-15).

*The Velsicol Company Dumpsite Near Toone, Tennessee**

Even when the chemicals are, supposedly, properly and legally collected and disposed of, there can be problems. For example, in the fall of 1978, residents in

*Much of this information was kindly supplied by Martha Teichner, CBS News, October 11, 1978.

FIGURE 16-15.   A hydrofluoric acid neutralization unit in San Francisco used to render some hazardous wastes safe for transport or amenable for recovery. (*Photo courtesy Chemical Manufacturer's Association.*)

Toone, Tennessee, particularly along the Toone-Teague Road, were experiencing fairly large quantities of various chemicals in their well water. The area of Toone affected was not large, only ten houses or so. About one-half mile from Toone, the Velsicol Company maintains a dumpsite of approximately 24 acres, although chemicals were buried (between 1964 and 1972) only in a 6-acre area. The chemicals were trucked to the site from the Velsicol Agricultural Chemical plant in Memphis, and (completely legally) buried in drums. Local residents who assisted or watched the disposal say that many of the drums broke open, spilling the chemicals.

Velsicol claims the dump site cannot be the source of the pollution, since the direction of the underground water flow is not between the dump site and the contaminated wells. On the other hand, all but one of the chemicals found in the contaminated well water occur at the dump site—substances such as naphthalene, chlorobenzene, toluene, tetrachloroethylene, and trimethyl benzene. There are no other dump sites nearby.

The people in this area of Toone were told not to drink the water and, if possible, not to bathe in it. Clean water was brought in tanks to the community. Eventually, the residents affected were able to tie into a water line being run from the rest of Toone, but they had to pay for the service, and construction took quite a while. Fears were that this water, from 5 miles away, would soon be contaminated also.

So far, the recent local health problems have been quite varied. One man experienced the symptoms of a stroke, but did not have a stroke. His daughter, a woman in her 30s, had numbness in her legs continually until she stopped drinking the water. Her sister-in-law delivered a premature baby with its inner organs outside its body. The child lived but is developing slowly. Most of the animals won't touch any water from local wells. One young dog did, and went blind and then died.

The issue is further complicated by the fact that the State of Tennessee has drilled test wells near the dump site. These wells penetrate into the deep subterranean passageway. There are fears that this could cause the chemicals to seep into the water supply for a large portion of the area.

FIGURE 16-16.   This wastewater treatment system at Shell Chemical Company's plant at Norco, La., cleanses four million gallons of water daily. The $35 million dollar complex uses biological oxidation to clean process water before discharge into the Mississippi River. (*Photo courtesy Chemical Manufacturers Association.*)

Velsicol and the state have joined together to study the question, with Velsicol footing half of the bill. However, Velsicol has no intention, at present, of paying to directly help the residents, or of relocating the 16.5 million gallons of chemical waste.[32]

The Baton Rouge Chemical Plant Treatment Facility[33]

An example of a modern wastewater treatment facility for chemical wastes is that installed at Exxon Chemical Company's Baton Rouge, Louisiana Plant. This facility, which went on-stream in May 1978, is designed to reduce organic discharge by 96% over 1968 levels.

There are six major steps in the treatment procedure.

1. The wastewater initially goes through primary separators and pretreatment to neutralize the pH.
2. The water then travels to dissolved air flotation units, where alum is added to act as a flocculent. An air-saturated water stream is recycled through these tanks to provide the air bubbles that cause the floc plus the collected oil and solids to float.
3. The "float" material is skimmed off. This sludge goes to a sludge dewatering section, where much of the water is removed in filter presses, and is then disposed of by landfilling.
4. The wastewater is next treated by trickling filters.
5. Additional microbial treatment occurs in a series of biological oxidation vessels or "reactors." Within these reactors, the microbes are supplied plenty of oxygen.
6. The wastewater stream lastly goes to clarifiers. The oxygen-activated sludge settles and is then recycled to the reactors. Any surplus is sent to an aerobic digester for stabilization and odor removal and then dewatered and disposed of in a landfill. The clarified water overflows into an adjacent bayou.

As can be seen, this procedure basically utilized the same techniques common to most other types of water treatment facilities. The major exception is, perhaps, the combined trickling filter/biological oxidation reactor microbe processing of the biodegradable materials. The plant itself cost $17 million, and it is designed to treat up to 5700 gallons of wastewater/minute.

Plastics Reduction

A large number of the chemical companies have responded to concerns about waste plastics and the effect on municipal landfills. There have been two approaches to this problem: 1) the development of degradable plastics, and 2) recycling of existing plastics, combined with product reduction.

Degradable Plastics[34,35]

Most commercial products made from degradable products are designed to either biodegrade or photodegrade. The majority of plastics are sensitive to ultraviolet (UV) light due to impurities that can initiate free radical reactions. The intentional addition of organometallic or metal compounds or the incorporation of photosensitive functional groups within the polymers can promote this degradation process.

Plastics are generally not biodegradable because the long, synthetic polymers cannot be digested by microorganisms. The addition of a few percent cornstarch, starch acetate, or cellulose acetate permits microorganisms to digest the starch, weakening and then breaking down the item.

Degradation time depends on the size of the article, especially the thickness; moisture, oxygen availability, and temperature for those that biodegrade; and exposure to UV light for those that are photosensitive. The degradable materials have been designed to be used in specific applications and disposed of under specified conditions. Major applications include six-pack beverage rings, trash bags, disposable diapers, compost bags, and grocery bags. A number of states (Florida, Nebraska, Rhode Island, West Virginia) either outlaw or tax nondegradable plastics for certain of these applications.

Plastics Minimization and Recycling[36,37]

Plastics recycling has included polyethylene terephthalate (PET), high-density polyethylene (HDPE), which jointly comprise 86% of all plastic bottles, and polyvinyl chloride (PVC), which makes up another 7%. Thermoplastics such as these are the most recyclable form of plastics because they can be remelted and reprocessed with only minor changes in their properties.

Typical processing includes a number of general stages:

1. Feeding, shredding, and prewashing;
2. Washing and separation;
3. Mechanical dewatering;
4. Thermal drying;
5. Storage; and
6. Extrusion after melting, with repelletizing.

In spite of significant public interest, only a few percent of plastics are recycled, mainly for "low-grade" products such as plastic lumber, park benches, landscaping materials, pallets, waterfront or shoreline construction components, and floor tiles. Some deterioration in quality inevitably occurs each time the polymers are reprocessed.

The plastics industry has simplified recycling by labeling containers with a code number, representative of the type of plastic.

- No. 1 is polyethylene terephthalate (PET), used often in plastic soda bottles, peanut butter jars, and similar containers. Because it is a polyester, it can be recycled to manufacture products such as fiberfill, polyester fabric, and carpeting. About 20% is recycled.
- No. 2 is high-density polyethylene (HDPE), which is used for approximately 65% of all plastic bottles (milk jugs, detergent bottles, and similar). A small amount (21%) is recycled for items such as trash cans, flower pots, and traffic barrier cones.
- No. 3 is polyvinyl chloride (PVC), used for flooring, credit cards, siding, and some packaging. Little is recycled.
- No. 4 is low-density polyethylene (LDPE), which is used for plastic bags, shrink wrap, and similar film wrappings. Many retail stores have established plastic bag recycling programs, but the overall amount recycled is negligible.
- No. 5 is polypropylene, used for aerosol caps, bottle lids, cottage cheese containers. Very little (~1%) is recycled.
- No. 6 are polystyrene and polystyrene foam, which are primarily used for food containers and packing "peanuts."
- No. 7 is mixed plastic, where several types are intimately combined in a particular product. Mixed plastic cannot be recycled due to the inability to separate the various types.

A number of large-scale plastics converters are approaching the problem from a waste minimization perspective. Procter and Gamble, for example, has taken a very aggressive approach to reducing the amount of plastics used in their products. These modifications include:[38]

1. Redesign of vegetable oil bottles to use 28% less plastic yet hold the same amount of oil;
2. Use of a polymeric sorbent material in diapers that reduces the volume by 50%, and squeeze the air from them prior to packing in plastic polybags rather than cardboard cartons, which overall reduces the packaging volume by 80%; and
3. Provide refills of cleaning products (dishwashing liquid, fabric softener, liquid laundry detergent) in concentrated pouches to be diluted by the user, reducing plastic consumption by 85%.

The switch from plastic packaging to paper and paper-based materials is epitomized by McDonald's decision in 1990 to drop its use of polystyrene foam clamshell-like boxes. It was estimated[39] that McDonald's used approximately 5% of the 905 million lb of polystyrene packaging produced in the United States in 1989. Estimates by the Environmental Defense Fund were that McDonald's could reduce its waste volume 90% by banning the packaging.

Controversy exists over which approach—degradable plastics or recycling—is better, and regarding whether plastics are really the solid waste culprit they have been accused of being. To a large extent, these arguments are both based on conjectures as to what really happens within landfills.

Recent studies, by boring up to 90 ft into four municipal landfills, have refuted some common assumptions as to the fate of various buried wastes.[40] Plastic fast-food packaging was found to occupy about 0.25% of the volume, whereas some estimates put it as high as 10%. Disposable diapers, estimated to occupy 5% of the volume, instead averaged about 1%. Overall, plastics items accounted for only 12% of the volume in a landfill, whereas some appraisals had put it at 30%.[41]

A major reason for the discrepancy appears to be that, after burial for a year or two under tons of refuse, dirt, and debris, anything plastic is squashed flat. Many glass containers, on the other hand, retain their shape.

The major materials in landfills are biodegradable wastes (especially newspapers and yard wastes), which typically occupy 65% of the volume, a percentage that is little changed after 10 years. The fear of some scientists is thus that "total" degradation of biodegradable plastic may require 20 years or more, and then will result in the pile of many little pieces, which will still occupy the same volume.

THE INORGANIC CHEMICALS INDUSTRY

The inorganic chemicals industry is quite diverse. Here we shall consider the production of only two inorganic chemicals, sulfuric acid (H_2SO_4) and ammonia (NH_3).

Sulfuric Acid

H_2SO_4 is the "top volume" chemical produced in the United States today. The majority of it is used to produce other chemicals, the greatest use (65%) being for the production of phosphate fertilizers. The mineral fluoropatite, $Ca_5(PO_4)_3F$, when reacted with H_2SO_4, forms a superphosphate, $CaH_4(PO_4)_2 \cdot H_2O$. This superphosphate is quite water-soluble; thus it is a good source of phosphorus for plants. Other uses for H_2SO_4 include use as the electrolyte in lead storage batteries; for pickling in steel production; in the production of organic dyes, plastics, and drugs; and in treating oils to de-emulsify them for wastewater treatment in refineries.

H_2SO_4 reacts strongly with H_2O, and thus is a good dehydrating agent. It can react to form SO_2 and H_2O, releasing an oxygen atom, and thus is a good oxidizing agent. It is also not very volatile, which is of great advantage in industrial applications.

At one time, H_2SO_4 was manufactured from sulfur deposits in Louisiana. Now the majority comes from byproduct sulfur from metallurgical plants, and from stack gas cleaning of power plants. Other possible sources include iron pyrite (FeS_2) and sulfide ores.

The majority of H_2SO_4 is made by the "contact process." Sulfur is first burned in air to produce SO_2: $S + O_2 \longrightarrow SO_2$. Next, SO_2 and O_2 are passed over a vanadium catalyst and react to produce SO_3:

$$2SO_2 + O_2 \xrightarrow{\text{vanadium oxide catalyst (V}_2\text{O}_5)} 2SO_3.$$

The SO_3 is then dissolved in H_2SO_4: $H_2SO_4 + SO_3 \rightarrow H_2S_2O_7$. When the pyrosulfuric acid or oleum ($H_2S_2O_7$) is mixed with H_2O, the reaction produces very concentrated (97%) H_2SO_4: $H_2S_2O_7 + H_2O \rightarrow 2H_2SO_4$.

In the production of H_2SO_4, as in some industrial processes that use H_2SO_4, the primary pollutants are emissions of SO_2, SO_3, and/or H_2SO_4 mist.

Ammonia

NH_3 the fifth highest volume chemical, is used predominantly in the manufacture of fertilizers. It can be applied directly to the soil as a gas or water solution, or it can be converted to ammonium nitrate, NH_4NO_3, ammonium sulfate, $(NH_4)_2SO_4$, ammonium phosphate, $(NH_4)_3PO_4$, or urea, $(NH_2)_2CO$. It is used, in addition, in the production of livestock feed, plastics, and other chemicals. It also is a good refrigeration gas due to its excellent heat transfer properties.

NH_3 is produced commercially by the Haber process. Nitrogen and hydrogen gases are reacted at high temperature and pressure: $3H_2 + H_2 \rightarrow 2NH_3$. The nitrogen gas is readily available from our atmosphere. The hydrogen is often obtained by reacting CH_4 and steam in the presence of a nickel catalyst:

$$CH_4 + 2H_2O \xrightarrow{\text{Ni}} CO_2 + 4H_2.$$

It is also available from coal, in a two-step process:

$$C(\text{coal}) + H_2O \longrightarrow CO + H_2$$
$$CO + H_2O \longrightarrow CO_2 + H_2.$$

NH_3 vapors, if breathed in too large a concentration, can be fatal; it is important to properly contain and/or vent the vapors.

Environmental Effects of Inorganic Chemicals

There are fewer environmental difficulties associated with inorganic chemicals production than with organic chemicals production. The majority of the emissions can be collected and disposed of by conventional techniques (see chapters 3, 4, and 5).

One major problem that still must be faced, however, is that associated with spills. Not only are oil spills damaging, but so are spills of H_2SO_4, NH_3, and other inorganic (or organic) substances. In recent years, much effort has been put forth in developing ways of handling these spills (other than simply evacuating the affected locality).

Heavy metal spills can be treated by sodium sulfide, leading to the formation of metal sulfide precipitates. This precipitate is readily visible, even if the spill is in a body of water, so a very small quantity of sodium sulfide can be used to locate the spill boundaries. Small spills can be treated this way; large spills require further investigation. The sulfide precipitates prevent the downward flow of liquids, and thus prevent groundwater contamination. This treatment is also feasible on land spills.

Most chemical spills on land can be immobilized by treatment with a gelling agent consisting of four different polymers, each to congeal one type of chemical, plus an inert powder, to improve the flow and dispersion characteristics. Once congealed, the spills can be collected by mechanical means. These gelling agents can also be used to seal splits in containers, thus minimizing the quantity spilled.

Many hazardous liquids can be immobilized by absorption in materials such as fly ash. H_2SO_4 is one such chemical. After absorption, the H_2SO_4 can be slowly neutralized by crushed limestone ($CaCO_3$).

Another possibility for those substances that are immiscible with and heavier than water is to collect the wastes in a temporary artificial sump area. The material can then be pumped to containers on the surface.[42]

Much work must still be done on chemicals recovery. Ruptured tank cars containing chlorine gas, NH_3, or similar substances can still create disasters, as we read all too often in our daily newspapers. Improved transportation regulations are only part of the answer.

References

1. "Facts & Figures for the Chemical Industry" *Chemical and Engineering News* **64**, June 18, 1990, pp. 34–83.
2. Borman, Stu, "Push for New Materials, Chemicals from Biomass Sparks Active R & D." *Chemical and Engineering News* **68**, September 10, 1990, pp. 19–22.
3. Davies, Duncan S., "The Changing Nature of Industrial Chemistry." *Chemical and Engineering News*, March 6, 1978, p. 22.
4. Fields, David, "Scientists Weed Out 'Energy Plants'." *Chicago Sun-Times*, August 25, 1978, p. 42.
5. "Some Trash Can Really Be Sweet." *Environmental Science and Technology* **9**, No. 12, November 1975, p. 1011.
6. *Oil*. Washington, D.C.: Energy Research and Development Administration, EDM-078 (5-77), pp. 1–2.
7. Pyle, James, *Chemistry and the Technological Backlash*. Englewood Cliffs, New Jersey: Prentice-Hall, 1974, p. 99.

8. Dedera, Don, "Disasters that Didn't." *Exxon USA* **XVI**, *No. 3*, p. 11.

9. Sell, N. J. and De Keyser, K. J., "Hazardous Spill Containment and Collection." *Pollution Engineering* **19**, *No. 7*, July 1987, pp. 74-78.

10. Sell, N. J., "Principles and Application of Sorbent Pillows for Collection of Hazardous Spills." *Environmental Management News* **1**, *No. 5*, May 1986, pp. 1-3.

11. "Oil Spills: An Environmental Threat." *Environmental Science and Technology* **4**, *No. 2*, February 1970, p. 97.

12. "Oil Spill Technology Makes Strides." *Environmental Science and Technology* **5**, *No. 8*, August 1971, p. 674.

13. "Cleaning Up Oil Spills Isn't Simple." *Environmental Science and Technology* **7**, *No. 5*, May 1973, p. 398.

14. *Jeanne Sorb: Revolutionary, Biodegradable Hydrocarbon Absorbent* J. V. Manufacturing Co, Inc., 1990, p. 2.

15. "A Tiny Solution to a Big Problem." *Chemecology* **19**, *No. 7*, September, 1990, pp. 2-3.

16. Ard, Lt. R. W., Jr., "Coast Guards' Response to Spilled Oil." *Environmental Science and Technology* **10**, *No. 3*, March 1976, p. 239.

17. "Counting on Bacteria by the Billions." *Span* **XVI**, *No. 3*, 1976, p. 16.

18. Grutsch, James F., "Wastewater Treatment: The Electrical Connection." *Environmental Science and Technology* **12**, *No. 9*, September 1978, p. 1022.

19. *Pollution Control Technology*. New York: Research and Education Association, 1973, p. 456.

20. Manahan, Stanley, *General Applied Chemistry*. Boston: William Grant Press, 1978, p. 394.

21. "Allied Has Hard Time Disposing of Kepone." *Chemical and Engineering News*, November 19, 1976, p. 6.

22. *Chemical and Engineering News*, March 14, 1977, p. 20.

23. Bertram, Carl G., "Minimizing Emissions from Vinyl Chloride Plants." *Environmental Science and Technology* **11**, *No. 9*, September 1977, p. 864.

24. "Are You Drinking Biorefractories Too?" *Environmental Science and Technology* **7**, *No. 1*, January 1973, p. 14.

25. *Pollution Control Technology*. p. 454.

26. "Chemical Industry Reduces Generation of Hazardous Waste." *Chemecology* **18**, *No. 8*, October 1989, p. 9.

27. Raloff, Janet, "Abandoned Dumps: A Chemical Legacy." *Science News* **115**, *No. 21*, May 26, 1979, p. 348.

28. "Chemical Industry Warned on Waste Dumps." *Chemical and Engineering News*, February 19, 1979, p. 6.

29. "Cleanup Goes Slowly in Chemical Warehouse." *Chemical and Engineering News*, May 21, 1979, p. 5.

30. Murray, Chris, "Chemical Waste Disposal a Costly Problem." *Chemical and Engineering News*, March 12, 1979, p. 12.

31. "Valley of the Drums and Other Hazardous Wastelands." *Science News* **115**, *No. 5*, February 3, 1979, p. 68.

32. "EPA Criticized for Chemical Wastes Handling." *Chemical and Engineering News*, November 13, 1978, p. 18.

33. Williamson, Alane G. and McDonnell, J. J., News Release, Exxon Chemical Company U.S.A., May 5, 1976.

34. Thayer, Ann M., "Degradable Plastics Generate Controversy in Solid Waste Issues." *Chemical and Engineering News*, **68**, June 25, 1990, pp. 7–14.

35. "Biodegradable Plastics: Friend or Foe of Waste Management." *Chemecology* **19**, *No. 8*, October 1990, pp. 2–3.

36. Basta, Nicholas and Johnson, Eric, "Plastics Recycling Picks Up Momentum." *Chemical Engineering* **96**, *No. 7*, July 1989, pp. 30–33.

37. Hanson, David. "Solid Waste Problem Receiving More Attention from Congress." *Chemical and Engineering News* **63**, July 3, 1989, pp. 23–24.

38. "Plastics and the Solid Waste Disposal Issue." *Chemecology* **19**, *No. 3*, April 1990, pp. 4–5.

39. "McDonald's to Drop Polystyrene Packaging." *Chemical and Engineering News* **64**, November 12, 1990, p. 5.

40. "Archaeologists Dig into the Secrets of Landfills." *Chemical and Engineering News* **63**, July 3, 1989, p. 60.

41. Thayer, Ann M., "Solid Waste Concerns Spur Plastic Recycling Efforts." *Chemical and Engineering News* **63**, January 30, 1989, pp. 7–15.

42. Dahn, Douglas and Pilie, Roland, "Technology for Managing Spills on Land and Water." *Environmental Science and Technology* **8**, *No. 13*, December 1974, p. 1076.

17

Impact of Environmental Regulations

The era of the environment can be dated in the United States from 1970, with the founding of the EPA. What has happened since then to environmental quality? How effective have the various regulations and cleanup programs been? What have been the costs? How could things be improved for the future? This chapter will briefly look at these and a number of other issues.

ENVIRONMENTAL QUALITY[1-4]

Air

Most U.S. cities are visibly cleaner now than in the early 1970s, cities such as Chattanooga, TN, Charleston, WV, and Grand Rapids, MI. Large particulates (smoke, soot) have decreased 61%; SO_2 emissions are 28% lower; there has been a continual reduction in Pb (96%) and CO (38%).[5] Why then did 81 cities recently still fail to meet clean air standards? The culprit is primarily the automobile, not industrial emissions. Ozone from automobile exhaust in particular has been a major problem in many urban areas.

But not all industries have "cleaned up their act." In 1990, residents in the northeastern part of Detroit suffered under a constant drizzle of oily soot, as a local steel plant continued to operate in apparent gross violation of the air pollution standards.

The far distant transport of air pollutants has only in later years been recognized as the major problem it is. In the Grand Canyon, the view is obscured by pollution 90% of the time. It is estimated that visibility in that region has decreased greater than 50% since 1950. Denver, which once was a refuge for people with illnesses such as tuberculosis, now has a death rate from lung disease some 30% greater than

FIGURE 17-1. A limestone plant located amongst the Canadian Rockies. No matter how far we get from civilization, we will always be faced with industry and its effects. How well the planet survives and thrives will be determined by many various key decisions we make in the next decade.

the national average. Airborne pollution from as far away as South America was proven to affect the Great Lakes.

In spite of significant reductions in SO_2 emissions, acid rain continues to take its toll. The German automaker BMW threatened to stop shipping cars through Jacksonville, Florida, due to acid showers that had purportedly ruined the paint on thousands of vehicles that were on the docks, awaiting transport. Moreover, unpredicted effects have been found. For example, acid rain has been correlated to supplying 25% of the nitrogen needed by the algae that are stifling the Chesapeake Bay. However, a 10-year study of acid rain found that, though widespread, its effects are not as bad as some scientists had predicted. Of the acidified lakes, only 4% are too acidic for the survival of any aquatic species; an additional 5% are too acidic for certain specific species.

Water

Overall, the nation's water quality has vastly improved since the 1970s. Specific success stories include the Susquehanna, Delaware, Willamette, and Potomac Rivers, and Lake Erie, which was once considered "dead." However, three categories of problems still exist: non-point runoff, the quality of drinking water, and the condition of coastal waters.

Many of the problems of coastal waters are directly caused by municipalities and only indirectly by industrial waste generators. The Clean Water Act of 1972 forbade the discharge of sewage unless > 85% of the bacteria and other pollutants are removed. This deadline was then extended to July 1, 1988. In spite of these deadlines, over 30 East Coast cities have continued to dump sewage into the ocean after only screening out the large, floating objects.

The disposal of sewage sludge into our oceans was also outlawed in 1981. However, New York was granted a court-ordered reprieve due to the expense of other alternatives. As a result of this dumping, the New Jersey coast no longer supports marine life. The disposal area was moved further from the shore in January 1989, initiating the daily dumping of 5.5 million gallons of sludge at the edge of the continental shelf. Congress mandated a total ban on such dumping, to take effect during the early 1990s.

Oil spills also continue to plague the coastal environment. Three major spills (in Rhode Island's Narragansett Bay, the Delaware River, and Texas's Galveston Bay) occurred in one horrendous, record-setting 12-hour period in June 1989. A coast guard study estimated that an average of 425 oil spills and 75 chemical spills occur monthly.

Oyster and other shellfish beds have been closed, or the harvests heavily reduced, along the coasts of Louisiana and Texas and in the Chesapeake Bay. San Francisco Bay is still heavily polluted; several Superfund sites have been identified in Puget Sound.

A number of different problems have threatened drinking water quality.

- The EPA has estimated that ⅔ of the 15,600 wastewater treatment plants in the United States have documented water quality or public health problems. To bring these plants up to standard would cost an estimated $83.5 billion.
- It is likewise estimated that ½ of the 5000 EPA-regulated landfill sites are leaking hazardous wastes into the environment.
- Streams in the Rocky Mountains, some 100 miles on both sides of the Continental Divide, are affected by mineral residues from some 10,000 abandoned mines. It is feared that these effects might pose a threat to surface water supplies throughout much of the West.
- Pesticides have percolated into well water in 34 states, significantly contaminating over 10% of the wells tested.

Non-point runoff from farms, mines, streets, sidewalks, and construction sites is responsible for a major share of the contaminants in our waterways. Even in Montana, usually considered to be fairly immune from most industrial pollutants, there are 4000 miles of streams that are not swimmable or fishable due to non-point pollution. Though efforts are underway to attack the non-point sources, control is much more difficult than for point source pollution.

Toxic Substances

The problem of toxic substances is becoming increasingly obvious. For example, some 320 airborne toxics have been identified, 60 of which are cancer-causing. The EPA continually adds toxic substances to their list of those substances controlled. Originally, 14 were listed to be regulated by RCRA guidelines—6 pesticides and 8 metals. In 1990, 25 additional organic chemicals were added, including some relatively common substances, such as benzene and carbon tetrachloride. It is estimated that the addition of these chemicals to the RCRA list will affect 17,000 additional sources of waste, in industries such as petroleum refining and marketing, pulp and paper, synthetic fibers, and organic chemicals.

Table 17-1 summarizes the amount of hazardous wastes produced in the United States in 1987, and their "ultimate" disposal method.

These numbers must be evaluated carefully, for they can be deceptive. A substance is considered hazardous even if only a small percentage is hazardous. The reported amount, therefore, may be many times the mass of the toxics present. For example, Fort Howard Corp. has been listed as Wisconsin's greatest generator of toxic air pollution, with more than half of the total (710,000 lb/year) consisting of hydrochloric acid (HCl) produced as a byproduct of coal combustion.[6] This HCl is extremely dilute, because coal contains only trace amounts of chlorine. However, that 710,000 lb is nonetheless the reported amount when toxic emissions are determined.

Toxic chemicals, however, have been identified in almost all locations.

- Traces of more than 400 toxic chemicals have been found in the Great Lakes, even after a $9 billion program had been implemented to control industrial and municipal effluents.

TABLE 17-1   Hazardous Wastes[a]

Route	Quantity (1000s tons)
Atmosphere	1350
Surface Water	4850
Underground Injection	1600
Municipal Facilities	950
Off-site Treatment/Disposal	1300
Landfill	1200
Total	11,250

[a]Data from Patterson, James W. "Industrial Wastes Reduction," ES&T, 23 (9), 1989.

- About 10% of the lakes, rivers, estuaries, and coastal waters contain sufficient toxics to be dangerous to aquatic life.
- Nine-tenths of the sport fish are tainted with toxics such as PCBs and DDT; one out of four fish have levels considered dangerous to humans who might ingest them.
- Water supplies in 75 major urban areas, including Washington, D.C., Baltimore, Miami, Cincinnati, and Philadelphia have been proven to contain chloroform and other carcinogens. New Orleans, at the mouth of the Mississippi River, is particularly hard hit due to all the polluters upstream.
- According to a 1990 EPA report, there were 15 chemical accidents in the United States in the 1980s with volumes and levels of toxicity of the chemicals greater than the 1984 catastrophe in Bhopal, India, where 3000 died from exposure to methyl isocyanate. No one was killed in the U.S. spills, primarily because most of the toxics involved were liquids.

Most current control methods simply transfer the toxic substances from one medium to another: toxic air contaminants are collected by wet scrubbers (and hence become a water pollution problem) or electrostatic precipitators (producing a solid waste problem). However, the old adage of one person's junk being someone else's treasure has also been realized in recent years with the development of "waste exchanges." Waste exchanges transfer either information about the availability or need for raw materials or about the raw materials themselves. Such exchanges have flourished and regularly transfer large quantities of waste between industries. These exchanges are located in numerous states, including California, Michigan, Indiana, Illinois, New York, New Jersey, Montana, North Carolina, and Florida.[7]

Solid Waste

Recycling has doubled in the United States since 1960, yet it accounts for only about 13% of the nation's solid waste. Landfill disposal accounts for 80% of our trash, twice that in most other countries.

Recycling is, however, probably the most impressive, brightest spot in U.S. society for the 1990s. Industries and individuals alike have exhibited considerable enthusiasm for recycling, sometimes more than can be exploited. Often, residents have carefully separated the various types of plastic and colors of glass, only to find their work went for naught, since it all ended up together in the landfill anyhow, due to a lack of markets for the recycled materials.

Massachusetts and Wisconsin were the first states to ban the disposal of a broad range of recyclable materials, from newspapers to plastic containers. In Massachusetts, landfill and incinerator operators can be fined $25,000/day if they improperly dispose of the specified recyclable substances.

The recovery of process wastes is also increasing. Increasing fuel prices and landfill disposal costs, let alone the political difficulty in siting new landfills, provide much incentive.

Overall Trends

Being, or at least sounding like, an environmentalist is finally good business and good politics. Environmental concerns have finally become part of the very structure of most U.S. institutions.

Strategies for reducing pollution at the source, rather than using end-of-pipe collection devices, are finally "in vogue" with both the EPA and corporate America. Though collection equipment has had some success, it always suffers from the problem of shifting the pollution from air to land to water and perhaps back again. Moreover, technology-based regulations have left many major problems, such as, potential global climatic change and possibly a thinning stratospheric ozone layer. Some of the regulations may have even been counterproductive, in that they have inhibited innovation and encouraged regulated industries to do only the minimum, legal requirements.

Even on a regional or local scale, many problems persist. Our technology is continually becoming more sophisticated; as a result, we are discovering problems never thought to exist. When, for example, evidence of dioxin was first found at U.S. paper mills, only two laboratories in the country could perform the required analyses. Concentrations are now measured in parts per quadrillion.

Serious contradictions still exist. There is evidence that the federal government is one of the country's worst polluters. According to the General Accounting Office, federal facilities are twice as likely to violate the Clean Water Act as private corporations. For example, authorities at the Rocky Mountains nuclear weapons plant allegedly dumped toxic chemicals into public drinking water, secretly incinerated a number of hazardous substances, and then falsified documents to protect themselves.

Enforcement of environmental regulations and concurrent other governmental activities are also not coordinated. Avtex Fibers, Inc., of Front Royal, VA, located on the Shenandoah River, was identified as the second largest source of air toxics in the United States. During 1988-1989, the company was convicted and fined for 1968 violations of its water discharge permit and 1921 violations of worker safety laws. It was later convicted and fined for contempt of court after 99 additional water discharge violations were found. Avtex was also fined for the willful violation of asbestos removal laws and the unlawful discharge of PCBs, and its officials were subject to a related criminal investigation. The company was then ordered to begin a $9 million cleanup of its old waste lagoons and the associated contaminated groundwater, which had been declared a Superfund site in 1984. In spite of all evidence that its personnel had utter contempt for our government and its laws and regulations, Avtex was still operating with a number of government contracts intact when it was finally shutdown.

The environmental quality, as well as people's perception about it, also varies with geographical location. According to the Institute for Southern Studies, located in North Carolina, the South is "the nation's biggest waste dump." On the other hand, William Ruckelshaus, former administrator of the EPA, commented, "I can close my eyes and listen to somebody start to talk about acid rain, and I can tell you within 100 miles where he or she lives. Westerners say they want acid rain cleaned up, but they won't pay for it. New Englanders say their lakes are poisoned and Ohio Valley states should pay for the cleanup. Ohio Valley states don't think there's a problem. And Southerners say, 'It's your problem. Don't bother us.'"[8]

COSTS

Compliance Costs

The monetary costs of pollution control vary from industry to industry and company to company, but no one would say that any of the costs are insignificant. An EPA report, "Environmental Investments: The Cost of a Clean Environment," based on a study that was mandated by both the Clean Air Act and the Clean Water Act, summarizes past overall compliance costs and projects those costs for the year 2000.[9] The study found that, in 1990, the United States spent $115 billion/year, or 2.1% of its gross national product (GNP) on pollution control. However, by the year 2000, this cost could increase to $185 billion/year (in 1990 dollars), or 2.8% of its GNP.

Although pollution control costs are continually increasing, they are doing so at a decreasing rate. Between 1972 and 1973, the rate of increase was 14%; in the mid-1980s, it was 6-8%; by the late 1990s, it is estimated to be about 3% per year.

Contrary to a common perception, the United States expenditures for pollution control appear to be proportionately greater than those of several Western European countries. The latest date that comparable data is available for is 1985. In that year, the United States spent 1.67% of its GNP on pollution control, West Germany spent 1.52%, the U.K. 1.25%, France 1.10%, and Norway 0.82%.

The cost of some specific types of pollutants and/or industries can also be considered:

- In the period 1972-1985, for example, industry spent $180 billion on air emission controls, with utilities paying $60 billion of this total. To install controls on new power plants cost $340-450/ton of coal capacity.

- It has been estimated by the EPA that the cost of compliance due to adding 25 organic chemicals to the list of RCRA-controlled hazardous substances costs industry $400 million annually, but will save $3.8 billion in damage to ground-water.[10]
- Pollution abatement for the chemical industry has increased from $898 million/year in 1973 to $3.28 billion/year in 1986. Table 17-2 categorizes the expenditures.

If the pollution control procedures a company utilizes are not sufficient, the companies have to pay for the damages. These costs also can be exorbitant.

- Conoco, the oil refining subsidiary of DuPont, had to buy 400 houses and residential lots near its Ponca City (OK) oil refinery for $18 million, or about $45,000 per property, due to a suit that claimed Conoco emitted hydrocarbons that contaminated the groundwater. In addition, it had to reimburse residents located further away an additional $5 million, as well as undertake to treat the contaminated water, at a still undetermined total cost.
- The U.S. government sought $19 million, or an average of $82,000 per home, from Occidental, the successor to Hooker Chemical Co., as reimbursement for cash advanced to buy 232 houses at Love Canal in Niagara Falls, NY, due to problems associated with the former hazardous waste dump site.
- The government also sought $33 million, or $84,000 per house, as reimbursement for the 393 properties bought in Times Beach, MO, because of the dioxin contamination.[11]
- USX agreed to pay $1.6 million in civil penalties and spend up to $32.5 million to upgrade wastewater treatment facilities, conduct contamination studies, and

TABLE 17-2.   Pollution Abatement Costs for the Chemical Industry ($ millions)[a]

	1973	1986
Capital expenditures		
Air	$164.4	197.8
Water	214.6	325.5
Solid waste	16.8	101.0
Operating costs		
Air	174.1	646.5
Water	247.6	1301.8
Solid waste	80.2	705.9

[a]Data from Hanson, David, "Administration Documents 20 Years of Environmental Improvement." *Chemical and Engineering News* **68**, June 18, 1990, p. 16.

perform cleanup operations arising from illegal dumping at its steel mill in Gary, IN.[12]

A major determining factor in environmental cleanup costs is the extent of emissions cleanup required. This appears to be true regardless of the type of pollutant and the cleanup method.

- About 18 million tons of SO_2 are emitted annually in the United States by industry and utility companies. It has been estimated[13] that to reduce this amount by one-third (or 6 million tons of SO_2) will cost $300/ton; the next 2 million tons reduction will cost $500/ton; the next 2 million tons will cost $900/ton; and the next 2 million tons, bringing the total reduction to two-thirds, or 12 million tons, will cost $1400/ton.
- The pulp and paper industry spent $30 billion between 1970 and 1978 to reduce their water pollution 95%. However, to reach 98% removal by 1984, as mandated by EPA, and hence to remove another 3% of the pollutants, required an estimated additional $4.8 billion.

What else makes a regulation particularly costly? An analysis by the Business Roundtable, an association consisting of the chief executive officers from 192 U.S. companies, led to the identification of nine attributes:[14]

1. Continuous monitoring
2. "Forcing" a new technology on-line
3. Capital intensity
4. Recurring costs
5. Retrofitting
6. Specific compliance action
7. Inadequate risk assessment
8. Engineering solutions
9. Changing requirements

Non-Monetary Costs

Costs other than monetary are also of significance. Switching to a lower-sulfur coal, for example, cost 25% of the 16,000 coal workers in Illinois their jobs. Construction of new industrial plants now requires increased lead time for planning and construction. For a large industrial project such as an oil refinery, petrochemical complex, pulp mill, steel mill, or smelter, the approval process will require a minimum of two years. And then it may be denied. For example, a petrochemical plant proposed by Dow Chemical for a few miles east of San Francisco was abandoned, after a $6 million investment, due to environmental difficulties. The Kaiparowits power plant, to be built in Utah, was similarly abandoned.

Settlement Money[15]

Since the 1970s, most federal environmental laws have permitted anyone to sue polluters for breaking the law. Private parties who sue are entitled to be reimbursed for their legal fees if the other party is found guilty, and any penalty payments are to go to the federal treasury.

In practice, few suits are filed by individuals or local citizens' groups; over-whelmingly, such action is filed by a few of the prominent national environmental groups, many by the Sierra Club. Nearly all are settled out of court, with the polluter agreeing to stop pollution, pay any government fines, and make charitable dona-tions to various environmental groups. As a result, this settlement money is a significant source of funding for such organizations, which usually are not directly involved in the specific suit.

As an example, in a case that made national prominence due to accusations of lack of accountability, Hercules Inc. had repeatedly violated its EPA discharge permit. As part of the consent decree filed in U.S. District Court, Hercules agreed to pay a $483,333 fine directly to the U.S. Treasury, and to donate $483,333 to the University of Medicine and Dentistry of New Jersey, $241,667 to the Rutgers University department of environmental resources, and $241,667 to the American Littoral Society, a group dedicated to the preservation of waterways.

Concerns about this policy abound.

1. There is no one to oversee the use of this money once it is distributed.
2. The taxpayers lose, since the money does not go into the state or federal treasury.
3. Industries can write it off as a tax-deductible donation, further avoiding taxes.

COST/BENEFIT ISSUES

Few people would argue against the fact that environmental regulations have been very beneficial for our society and our lifestyles. However, few would disagree that they also have been very costly. The major issue, therefore, is how to get a "bigger bang for our bucks."

There is usually little debate over regulations when the harmful effects of a pollutant are large, immediate, and traceable to the substance in question, or even when there is a question about a substance that can easily be removed from an effluent by a collection technique or a process change. The difficulties and dilem-mas are apparent when a highly beneficial activity leads to low pollution levels of substances that have not proven harmful in epidemiological studies and that are very costly to eliminate. Then it becomes a question of risk analysis. Unfortunately, attempts to eliminate risks for a few often lead to greater risks—of a different type—for the remainder of the population.

Perhaps one of the first considerations when one looks at the environmental regulations is whether or not the regulation is based on "good science." There are continuing instances where this may not be true.

Dioxin, the chemical that forced the evacuation of Love Canal and became notorious as the most potent carcinogen ever tested, may be instead the most maligned chemical.[16] "New" scientific evidence has shown that it is probably far less dangerous than once imagined.

The case against dioxin dates back to two rodent studies conducted in the 1970s, both of which showed the chemical to be highly carcinogenic. However, since these tests were conducted, the scientific standards for evaluating a chemical's carcinogenic potency have become much more precise and reliable.

In 1989, Robert Squire, who performed the original analysis of the animal data for the EPA, reexamined his findings, using the original plates. His conclusion, endorsed subsequently by a panel of prominent pathologists, was that as many as half of the lesions originally thought to be evidence of dioxin's carcinogenicity were actually benign. The previous study has indicated that it is dangerous in even the smallest amount; the new assessment found that it caused no tumors except at the highest doses, making it only a weak carcinogen. Extrapolated to human exposure, this means that the daily safe dose of dioxin could be increased thirtyfold.

Classification as a weak carcinogen, at most, correlates with an epidemiological study conducted by the Center for Disease Control in Atlanta, GA. The Givaudun 2, 4, 5-trichlorophenol plant in Meda, Italy suffered a serious explosion on July 10, 1976. The travelling cloud deposited almost 3 lb of dioxin over 700 acres, including the city of Seveso. The population has apparently suffered no adverse health effects as a result of this exposure other than chloracne, a skin condition characterized by a bumpy rash, blackheads, and pus-filled blisters.[17]

A second (simultaneous) issue is the manner in which the EPA determines its exposure limits. Dioxin is ubiquitous in our environment: it is a byproduct of incineration, wood and coal burning, petroleum refining, chemicals, and pulp and paper manufacture. The daily exposure to Americans is estimated to be probably 0.5 picograms (0.5×10^{-12} g) per kilogram of body weight daily. The EPA has assumed that the maximum dose people should receive—for a 1 in 1 million chance to get cancer—is only 0.006 picograms per kilogram, or only about 1% of the dose people actually receive. To calculate this dose, the EPA has taken the original results of the rat study and directly extrapolated them to people, with the assumption of no threshold level.

Many scientists believe, instead, that below a certain level dioxin is safe. Canada, Australia, and a number of European countries have estimated that dioxin remains safe at doses from 1 to 10 picograms per kilogram per day, well above the normal exposure levels and several hundred times higher than the EPA estimate.

This difference in allowable exposure could have a significant economic impact. It has been estimated that upwards of $1 billion could be saved in the

United States by moving to a standard of 5 picograms. Even a standard of 1 picogram would eliminate 75–90% of the cleanup costs for sites contaminated primarily by dioxin.[18]

Similar concerns have arisen regarding the control of trichloroethylene (TCE).[19] TCE has been used since the 1940s, primarily as a dry cleaning solvent, a degreasing agent, an extraction agent in decaffeinating coffee, a general anaesthetic in medicine and dentistry, and in many other applications.

In 1976, a study by the National Cancer Institute indicated that, in very high doses, TCE caused tumor growths in one particular species of mouse. Studies by subsequent researchers have failed to reproduce this response in any other species of test animal, and epidemiological studies have likewise failed to implicate TCE as a cause of cancer in humans.

Though no direct evidence ever existed indicating that the ingestion of even small amounts of TCE was carcinogenic, in 1976 TCE was included on the EPA list of hazardous substances. In subsequent years, TCE has been found in groundwater throughout the United States, leading to hundreds of environmental litigations, and extensive environmental regulation. The EPA, for example, shut down 35 wells in California's Silicon Valley, due to TCE pollution. Tort litigations involving TCE are in the billions of dollars; remedial actions involving groundwater and soil cleanup to essentially nondetectable levels are costing more billions of dollars; and the populace has perhaps been unduly traumatized due to the exaggerated publicity and unwarranted allegations regarding its deleterious health effects.

The EPA itself admits a lack of proper priorities. The EPA Science Advisory Board stated that federal laws and programs "are more reflective of public perceptions of risk than of scientific understanding of risk."[20] As the toxicity of more and more compounds becomes a question, the classification of potential carcinogens and the establishment of permissible dose rates will become increasingly important. Over 50% of all chemicals tested to date have been found to be carcinogenic in high doses, so the problem must be addressed.

In the realm of water pollution control, various governmental agencies (including the EPA and Corps of Engineers), municipal treatment facilities, and industry groups have formed a new initiative, Water Quality 2000. This group is striving to identify concerns and problems to be addressed by upcoming Clean Water legislation. The participants are trying to reach broad concensus as what should be the guidelines for drinking water standards, non-point pollution, and so forth, so they can make recommendations to Congress. In this way, it is hoped that the guidelines can be both realistic and based on "good science."

Another policy change that industry personnel feel would be beneficial would be for the government to attempt to reduce the uncertainty that now exists about the future direction and the timing of various regulations. This continues to be a considerable problem. For example, a number of years ago the Pennsylvania Power Company, a subdivision of Ohio Edison, spent two years and $2 million installing air pollution control equipment to meet the 98% pollutant removal efficiency requirement established by Pennsylvania Air Pollution Regulation No. 4. No sooner was the system installed, when Regulation No. 5, requiring 99% efficiency, was issued. To remove that extra 1%, the newly completed system had to be scrapped, and an additional $4 million spent for a more efficient system.

SUMMARY

It is going to cost increasingly more in the future to further improve our environment. It will not be a simple job. It will create some undesirable effects as well as beneficial effects. Yet when we look at the progress that has already been made and at the tasks still facing us, it seems worthwhile to continue—with logic, knowledge, and good sense.

References

1. "The Year of the Deal." *National Wildlife* **29**, *No. 2*, February–March 1991, pp. 33–40.
2. "The Year of Broken Promises." *National Wildlife* **28**, *No. 2*, February–March 1990, pp. 45–52.
3. "The Planet Strikes Back." *National Wildlife* **27**, *No. 2*, February–March 1989, pp. 33–40.
4. "Things are Getting Better... and Worse at the Same Time." *National Wildlife* **26**, *No. 2*, February–March 1988, pp. 38–45.
5. Hanson, David, "Administration Documents 20 Years of Environmental Improvement." *Chemical and Engineering News* **68**, June 18, 1990, pp. 15–17.
6. "Fort Howard Tops List of Toxic Air Polluters in State in 1989." *Green Bay Press-Gazette*, February 4, 1991, p. A-1.
7. Patterson, James W., "Industrial Wastes Reduction." *ES&T* **23**, *No. 9*, 1989 pp. 1032–1038.
8. "The Politics of Acid Rain." *Contact* **57**, *No. 7*, July 1985, pp. 12–13.
9. Ember, Lois R., "Rising Pollution Control Costs May Alter EPA's Regulatory Direction." *Chemical and Engineering News* **69**, February 18, 1991, pp. 25–26.
10. Hanson, David, "Hazardous Wastes: EPA Adds 25 Organics to RCRA List." *Chemical and Engineering News* **68**, March 12, 1990, p. 4.
11. Reisch, Mark, "Conoco to Settle Toxic Suit by Buying Out Homes." *Chemical and Engineering News* **68**, April 9, 1990, p. 7.
12. "Government Roundup." *Chemical and Engineering News* **68**, August 6, 1990, p. 18.
13. "Q&A: An Acid Rain Primer." *Contact*, **57**, *No 7*, July 1985, p. 10.
14. "Cost of Federal Rules to Firms Detailed." *Chemical and Engineering News* **57**, March 26, 1979, p. 6.
15. Felton, Eric, "Money from Pollution Suits Can Follow a Winding Course." *Insight* **7**, *No. 7*, February 18, 1991, pp. 18–20.
16. Gladwell, Malcolm, "Scientists Temper Views on Cancer-Causing Potential of Dioxin." *The Washington Post*, May 31, 1990, A3.

17. Stinson, Stephen, "Paper Mill Wastes Show Low Dioxin Levels." *Chemical and Engineering News* **67**, July 24, 1989, p. 28.
18. Gladwell.
19. Schaumburg, Frank D. "Banning Trichloroethylene: Responsible Reaction or Overkill?" *ES&T* **24**, *No. 1*, 1990, pp. 17–22.
20. Ember.

Index